A PHOTOGRAPHIC

SN

OTHER REPTILES AND
AMPHIBIANS OF EAST AFRICA

BILL BRANCH

DEDICATION

To the Kihansi Spray Toad *Nectophrynoides asperginis* – a small victim of development. May we learn to value all wildlife.

Published by Struik Nature
(an imprint of Random House Struik (Pty) Ltd)
Reg. No. 1966/003153/07
The Estuaries No. 4, Oxbow Crescent, Century Avenue
Century City, 7441
PO Box 1144, Cape Town, 8000 South Africa

www.randomstruik.co.za

First edition published in 2005
This edition published in 2014

10 9 8 7 6 5 4 3 2 1

Publishing manager: Pippa Parker
Managing editor: Helen de Villiers
Editor: Emily Bowles
Designer: Neil Bester
Reproduction by Hirt & Carter Cape (Pty) Ltd
Printed and bound by Replika Press Pvt Ltd

ISBN 978 1 77584 165 4
ePUB 978 1 77584 257 6
ePDF 978 1 77584 258 3

ACKNOWLEDGEMENTS

I am grateful to the following colleagues for letting me use their photographs: Kenny Babilon, Wolfgang Böhme, Marius Burger, Paul Freed, Chris Kelly, Warren Klein, Reto Kuster, Simon Loader, Johan Marais, Michele Menagon, Mark-Olivier Rödel, Steve Spawls, Colin Tilbury, Phillip Wagner and Wolfgang Wurster. Emily Bowles was an able editor, and maintained consistency, style and deadlines. Joe Beraducchi (BMT, Arusha), allowed his rare captives to be photographed. I thank Steve Spawls, whose whose knowledge of the biology and distribution of East African reptiles made this guide so much more accurate.

Front cover photograph: Rufous beaked snake *Rhamphiophis rostratus*
Back cover photograph: Yellow-spotted tree frog *Leptopelis flavomaculatus*
Title page photograph: Flap-necked chameleon *Chamaeleo dilepis*

CONTENTS

KEY TO SYMBOLS USED IN THIS BOOK

 Snakes whose bite can be fatal to humans

 Venomous snakes whose bite is not known to be fatal

INTRODUCTION

This book introduces 274 of the snakes, lizards, crocodiles, chelonians and amphibians that live, often unnoticed, all around us. East Africa includes here the countries of Kenya, Uganda, Rwanda, Burundi and Tanzania, and has a combined area of approximately 1,819,000km^2 (about 20% of the size of the United States of America). It is host to a tremendous variety of habitats that harbour a magnificently diverse reptile and amphibian fauna. Over 450 reptile species have been recorded from East Africa, as well as more than 230 amphibians. Many of these are endemic, particularly to the isolated mountain regions, such as the Eastern Arc Mountains of Tanzania, which are a global hot spot of biodiversity. Unfortunately, not all species can be included in this small guide, and those chosen emphasise the more colourful and conspicuous species, as well as those unique to, common or endangered in East Africa. It is to be hoped that this small book will help to enrich wildlife experience and will convince readers that this wonderful thing called 'life' must be preserved in all its fabulous diversity.

Living reptiles are extremely diverse and the relationships between the main orders and families are still controversial. Three of the four surviving lineages occur in East Africa, including the crocodilians (Order Crocodylia); tortoises, terrapins and turtles (collectively called chelonians, Order Testudines); and the scaled reptiles include snakes and lizards (collectively called squamates, Order Squamata). Crocodiles are more closely related to dinosaurs and birds than to other reptiles, and chelonians have little similarity to other living reptiles. Among scaled reptiles, lizards are the ancestral group from which snakes and worm lizards (also called amphisbaenids) evolved. Limb loss in lizards has evolved many times, and it is not always easy to distinguish legless lizards from snakes. Most snakes have enlarged belly scales (ventrals) that aid locomotion, whereas legless lizards do not. In addition, the eyes of snakes lack eyelids and they have an unblinking stare. However, many snakes that burrow have evolved to be eyeless and without enlarged belly scales. Burrowing lizards may also become blind, and look very similar to burrowing snakes. There is no simple rule for telling them apart, but all blind snakes have very short tails with a sharp spine at the tip, and their heads are

Lizards, like this agama, are part of the ancestral group from which snakes and worm lizards descend.

rounder and blunter than those of legless lizards.

Reptiles have only limited mobility and many have very specific habitat requirements. In general, lizards can be considered habitat-linked, while snakes can be considered food-linked. This is reflected in many of their common names. Thus, among snakes, we talk of egg eaters, centipede eaters and slug eaters, and among lizards, of desert lizards, rock lizards and water monitors. In practical terms, this means that lizards often have very small ranges within which they inhabit specific places.

Snakes may range over large areas as they search for prey. Lizards, however, often inhabit very specific places.

Snakes often range over large areas and occur in different habitats, but search for specific prey.

The majority of reptiles, including all crocodiles and chelonians, are oviparous (egg-laying). Most lay clutches of between five and 20 eggs, although large sea turtles may lay up to 1,000 eggs in a season. By contrast, most geckos lay only two eggs at a time. With few exceptions, parental care in all local reptiles ends when the eggs are laid and the nest sites are covered. The exceptions are pythons and crocodiles, which both attend their eggs until they hatch. A few other snakes also stay with their eggs after laying them, although this is poorly studied. In crocodiles and many chelonians, the sex of the embryo is dependent upon the temperature at which the egg is incubated. In crocodiles, males develop in eggs at high temperatures, while the same temperatures in chelonians usually produce females. Many snakes and lizards are viviparous, retaining their eggs within the body and giving birth to live young. This is usually associated with life in cool climates.

Most African amphibians are anurans (Anura) that have well-developed legs and lack tails as adults. Tailed amphibians (Caudata: newts and salamanders) are absent in sub-Saharan Africa, although common in the northern hemisphere. The rare Gymnophiona (caecilians) are legless amphibians with rubbery bodies that live underground in damp soil. They occur in the tropical regions of South America, Asia and Africa, with few African (20) or East African (10) species. Amphibians are characterised by having a larval stage (the tadpole or pollywog) that is usually free-swimming. Some species, however, lay large yolky eggs in damp soil where the tadpole grows within the egg, emerging as a miniature froglet. In some frogs the eggs are even retained within the mother, who then gives birth to live young.

HABITATS

East Africa has a tremendous range of habitats, ranging from tropical beaches and coral reefs, moist (woodland) and dry (bushland) savanna, lowland rainforests and swamps, Afromontane forests and grasslands, to semi-arid scrub and rocky deserts. The map below shows the major vegetation zones in the region. The principal isolated uplands and mountains, such as Mts Kenya, Elgon and the Aberdare Mountains of Kenya, and Mts Kilimanjaro, Meru and the Eastern Arc Mountains of Tanzania, provide diverse habitats. Depending on their heights, they may have patches of montane forest, mountain heathland, and even alpine vegetation. Many of these contain endemic reptiles and amphibians. The suitability of these habitats also depends on climate, particularly rainfall, and reptiles and amphibians may only be seasonally active after rains, hiding away during the dry season.

Desert	Swamp
Savanna grassland	Dry savanna
Moist savanna	Grassland
Forest	Semi-desert

Left: *Dry savanna in the Nguila Hills, Tsavo West, Kenya in the wet season*
Right: *The same area of dry savanna in the wet season*

Left: *Moist savanna ('short miombo woodland') in southern Tanzania*
Right: *Coastal forest and grassland mosaic in southern Kenya*

IDENTIFICATION IN THE FIELD

Snakes

It is not difficult to distinguish the main groups of snakes in the field. Differences in coloration, behaviour and habitat all help to make a general identification possible. Precise identification may require the capture of the specimen for detailed scale counts. The main groups can be distinguished by the following general features.

Burrowing primitive snakes (pages 11–14)
Head blunt, with vestigial eyes; body shiny, and tail short, with a spine at the tip; belly scales not enlarged. Usually plain black or brown in colour, sometimes blotched. Burrow underground; wriggle but never bite.

Pythons (pages 15–16)
Head triangular, covered with small scales, and with prominent pits on the lips; body strong and muscular with smooth, small scales; belly scales enlarged; tail moderately long. Usually blotched in brown, cream and olive.

Adders and vipers (pages 17–25)

Head triangular, covered with small scales (with the exception of night adders), and usually with a prominent mark ('V'-shaped or arrow-shaped) on the crown; body fat, with rough small scales; belly scales enlarged; tail very short. Usually blotched in browns and greys. Active in the early morning or evening. Usually give a warning hiss, and strike readily.

Cobras and mambas (pages 26–34)

Head covered with large, symmetrical scales on crown; belly scales enlarged; large front fangs. Body form, coloration and habits very varied. Except for garter snakes, all threaten by raising the forebody and inflating or spreading the neck.

African and Old World snakes (pages 35–74)

Head covered with large, symmetrical scales on crown and enlarged belly scales. Body form, coloration and habits are varied.

Lizards

Lizards cause the most confusion, but they do have family characteristics that allow general identification.

Skinks (pages 75–83)

Skinks have smooth, shiny, cylindrical bodies, with narrow, pointed heads, no obvious necks, and tapering tails that can be shed and regrown. They move slowly, searching for food, and when disturbed usually slink behind cover, while keeping a watchful eye. Many burrowing species have short tails, and usually have no limbs, or vestigial limbs.

Old World lizards (pages 84–89)

Lacertids have similar body shapes to skinks, but their body scales are granular and not shiny, and the tail is usually noticeably longer than the body. It is shed easily and regrown quickly. They are mostly terrestrial, 'sit-and-wait' predators. When disturbed they sprint rapidly from bush to bush.

Plated lizards (pages 90–91)

Large terrestrial lizards that have shiny, cylindrical bodies. The body scales are rectangular and not rounded. A prominent lateral fold on the side of the body further distinguishes them from skinks. The long tail can be shed, but is regenerated quickly. They move around slowly, searching for food, which is often uncovered by scratching with the forelimbs.

Girdled lizards (page 92–93)

They have flattened bodies, girdled in rings of spiny scales. The head is triangular with a narrow neck and large, symmetrical scales on the crown. The tail is very spiny, and can be shed and regenerated. They bask, head up, on prominent rocks, and retreat into rock cracks or hollows when disturbed.

Monitors (page 94)

Very large lizards; even as hatchlings they are bigger than most other lizards. The snake-like tongue is constantly flicked in and out. The crown of the head is covered in small scales. The tail cannot be shed or regrown, and is used as a whip in defence.

Agamas (pages 95–98)

Medium-sized lizards with fat, spiny bodies, narrow necks, and well-developed legs. The tail cannot be shed or regrown. Often perch in prominent spots with the head held high. Males are very colourful.

Chameleons (pages 99–108)

Slow-moving, mainly arboreal species. Their unusual clasping feet, telescopic tongue and protruding eyes that move independently are all unique. The tapering tail cannot be shed or regrown. Males often have complicated crests and horns on the head.

Geckos (pages 109–117)

Geckos are mostly nocturnal, although day geckos, like other lizards, are active during the day. All climb on rocks or trees and can leap across small gaps. To aid this they usually have flared toe-tips. When disturbed they hide in rock cracks or behind branches. The head and body scales are small, granular and not shiny. The tail can be shed, but is regrown.

Worm lizards (page 118)

Worm-like, legless, with soft, usually pink-violet skin, with rings of soft, square scales, and vestigial eyes. Wriggle underground.

Crocodiles (pages 118–119)

Crocodiles look superficially like lizards, but have twin keels on the tail, which is squarish in cross-section. In addition, their hindfeet are webbed, and their nostrils and eyes are displaced to the top of the head. All are aquatic and never found far from water.

Chelonians (pages 120–127)

Tortoises, terrapins and turtles are characterised by their protective shells. They have different habits, and are distinguished by their feet. Tortoises live on land, and have thick, short feet. Terrapins live in fresh water, and have a webbed frill to the hindlimbs. Turtles are marine and the forelimbs are flippers.

Amphibians (pages 128–153)

Frogs are difficult to place in family groups as many of their key features are internal, and many unrelated species may look superficially similar.

HOW TO USE THIS BOOK

Clarity and ease of use have been the main criteria in the design of this photo guide. In the succinct species descriptions, key features for identification are emphasised in italic type. In some difficult groups, such as thread snakes and sand lizards, it may be necessary to have the specimen in hand to obtain detailed scale counts for identification.

As with birds, field identification is not always easy and depends on sensible and good observation.

> When first spotting a reptile, try to note the following features: What was the general build? Was the coloration uniform or patterned, and, if the latter, was it plain, striped or blotched? Was the tail longer or shorter than the body? Were the body scales large or small, smooth or rough, arranged in rows or scattered? What habitat was it in, and what was it doing? For frogs, judge coloration, the length of the head and hindlegs, behaviour, and habitat. These details should be jotted down in a field book, as an aid for comparison with the pictures in the species accounts.

The thumbnail silhouettes allow quick access to the reptile or amphibian group most closely resembling the species you have seen. A glossary of specialised terms and diagrams showing the anatomy of typical reptiles and amphibians is featured on pages 154–156. When finding an illustration that is similar to the unknown species, check the map to see whether it occurs in the region. If there is a big difference between where you have found it and where it should be, try another picture. Although many animal distributions are poorly known, you are unlikely to discover major range extensions. It is more likely that you have simply misidentified the species.

Remember that many species, particularly frogs and lizards, are not included in this guide. You are directed to a selection of further reading (p. 156) that should allow for the identification of any unusual or less common species.

A WORD OF CAUTION: The small maps used in this guide can only display the general area in which a species occurs. Within this range, however, the species is restricted to a suitable habitat. Crocodiles, water terrapins, and many amphibians and snakes are obviously dependent upon water, while other species are restricted to forest habitats or live on rock outcrops. Such habitats are fragmented and the species specific to them occur in isolated populations. As an additional complication, the distributions of many reptiles and amphibians in East Africa remain poorly documented. The maps for most reptiles show fragmented populations. Some of the gaps are created where the species is genuinely absent. Others, however, are areas that remain unsurveyed and where the species may still occur. For wide-ranging amphibian species only very general distributions are given and they can be expected to be absent from large areas within the shown range.

SNAKES Suborder Serpentes

BLIND SNAKES Family Typhlopidae

Primitive snakes that *lack teeth in lower jaw*. A blind snake's body is cylindrical, ringed in small, smooth, overlapping scales. A small, light-sensitive, black spot beneath the head scales is all that remains of *vestigial eye*. Snout rounded, sometimes with horizontal edge for pushing through soil. Has *very short tail* ending in spine. Blind snakes spend their lives burrowing underground, searching for brood chambers in ant nests where they 'binge' on eggs and larvae. Occur in tropical regions. Family has at least 20 local species.

ZAMBEZI BLIND SNAKE
Afrotyphlops mucroso 60–80cm

Top: *Blotched phase*
Above: *Uniform phase*

The largest blind snake in the world. Thick-bodied and distinguished by having *30–38 scales at midbody* and *prominent horizontal edge to snout*. Coloration variable, body may be plain or blotched. When skin is shed, body is bright blue-grey with dark markings, but skin tans with time to red-brown, with yellow on the belly. Locally restricted to southern Tanzania, extending along the coast to Kenya. Very large specimens seen only when forced to the surface by floods. Lives deep underground, crawling into brood chambers of ant nests and eating eggs and larvae. Large fat stores allow it to undertake long fasts. May feed only 2–3 times per year. Lays large numbers of eggs (12–40, and up to 60 in very large females) that take 5–6 weeks to hatch.

LINED BLIND SNAKE
Afrotyphlops lineolatus 58–64cm

A large, stout species with broad, rounded snout and wide rostral (¾ the width of the head). Coloration varied. Usually blackish, each scale with yellow spots giving a lined appearance, but sometimes with series of irregular pale blotches. Belly plain grey-white. Found in savannas and farm bush of West Africa, east to Ethiopia and northern Tanzania. Little known of its biology, but probably similar to that of other large blind snakes.

SPOTTED BLIND SNAKE
Afrotyphlops punctatus 40–66cm

A stocky blind snake distinguished by *30–32 scales at midbody* and *rounded snout* without horizontal cutting edge. Coloration variable, and dark brown or grey body may be speckled, each scale with a yellow spot. In some specimens, yellowish body is *heavily blotched, even on belly*. Restricted to northwest of the region, extending from Uganda to West Africa. Another forest inhabitant whose life history is poorly known, but probably similar to that of Zambezi Blind Snake (p.11).

ANGOLA BLIND SNAKE
Afrotyphlops angolensis 40–62cm

Large, long-bodied blind snake with rounded snout and only 28–32 scales at midbody. Body appears finely speckled, being rich grey-brown with pale-centred body scales. Belly unpatterned, cream-white to light brown. Locally restricted to western Uganda and adjacent regions, with possible isolated population in central Kenya. Extends also into Cameroon and Angola. Found in varied habitats from lowland forest to savanna and even montane grassland. Little known of its life history, but is probably similar to that of Zambezi Blind Snake (p.11).

FLOWER-POT SNAKE
Indotyphlops braminus 14–16cm

Very small, slender blind snake with *rounded snout, 20 midbody scale rows* and *300–350 scales along backbone.* Uniform grey to pale brown, with lighter belly and *cream blotches on snout and anal region.* An Asian

species, probably introduced to Mombasa early in last century, but now found in urban gardens along coast. A *self-fertilising, all-female species* that lays 2–6 minute eggs. Commonly transported in nursery plants, hence its common name, and has been carried to scattered spots around the world, including Cape Town, Florida and Israel.

THREAD SNAKES Family Leptotyphlopidae

Primitive snakes – the smallest in the world. Have very thin, cylindrical bodies covered in small, smooth scales. *Lack enlarged belly scales* and have *vestigial eyes*. *Lack teeth in upper jaw*. All local species have a blunt head and short tail that is relatively longer than that of a blind snake. Harmless and live underground, following the chemical trails of ants to their nests. Most species only distinguished with aid of microscope. Lay a few elongate eggs joined together like strings of sausages.

SUDAN BEAKED THREAD SNAKE
Myriopholis macrorhyncha 14–16cm

A small, elongate species with cylindrical body, slightly broadened head and neck, and *moderately long tail* (26–43 subcaudals) that tapers to small, blunt cone. Snout with distinct beak and body with 14 rows of smooth scales. Body pale reddish-brown or pink with cream to white belly. Found in sandy dry savanna and semi-desert, from northern Tanzania to Sahel and along Nile to the Levant and Turkey.

MERKER'S THREAD SNAKE
Leptotyphlops merkeri 4–9cm

A large thread snake with *rounded snout* and narrower, *wedge-shaped rostral*. Body cylindrical with

14 scale rows. Short tail (18–30 subcaudals) has only 12 scale rows and tapers abruptly to small terminal spine. Body uniform dark brown to black above and below, often with white patches on lower lip, chin and throat. Found in savanna from highlands of southeastern Kenya into eastern Tanzania.

PYTHONS Family Pythonidae

A primitive family with some of world's largest snakes, including Africa's largest, the Central African Python. There are four African species. Can give painful bites and large specimens should be treated with care. Human fatalities by constriction known, but very rare.

CENTRAL AFRICAN PYTHON ☠
Python sebae 350–750cm

A 9.8m python was reported from Ivory Coast, but lengths over 6.5m are very rare. Stout with small, smooth scales in *76–99 midbody rows*. Belly scales narrower than body. *Triangular head with medium–large scales. Upper lips have two heat-sensitive pits on each side. Large, dark spearhead mark on* crown. A broad band passes through each eye. Body blotched. Favours forest in Congo Basin, with scattered records in Kenya and Tanzania. Juveniles eat small mammals, while adults may tackle small buck. Female broods up to 100 eggs the size of oranges.

SOUTHERN AFRICAN PYTHON ☠
Python natalensis 350–580cm

Left: *Juvenile* Right: *Adult with prey in stomach*

Africa's second largest snake. Unmistakable stout body with very *small, smooth scales in 71–95 midbody rows. Triangular head* has *small irregular scales. Upper lips have paired heat-sensitive pits* on each side. A narrow band passes through each eye. Body blotched with *dark spearhead mark on crown.* Extensive in southern savannas. Favours rocky or bushy areas near water. Juveniles eat small mammals and ground birds, and adults may tackle small buck. Female broods up to to 80 eggs the size of oranges.

ROYAL PYTHON
Python regius 80–160cm

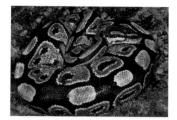

Small, *pear-shaped* head. Crown has *small fragmented scales and a dark triangle* with light sides. Prominent dark stripe extends from snout, through eye. Body dirty gold with irregular light-edged dark blotches. Stout with *53–63 midbody rows.* Extends through the Sahel, with a

few records in far western Uganda. Gentle and rarely bites but often rolls up, hiding its head among its coils. Inhabits dry and moist savanna, living in holes and feeding on rodents and birds. Can undertake long fasts during dry season. Lays 3–11 (usually 4–6) large eggs, which female incubates by 'shivering' to speed development. Very popular in the pet trade.

BOAS Family Boidae

A primitive viviparous snake family that includes the world's largest snake, the anaconda (*Eunectes murinus*) of the Amazon. The only African representatives belong to the Erycinae – small, burrowing boas restricted to the Old World.

KENYAN SAND BOA
Eryx colubrinus 40–90cm

A small, stout snake with boldly blotched *red-ochre* body. *Short head* covered above in *small scales,* and snout has *broad rostral* used like a bulldozer when burrowing. *Minute eyes* set near *top of head* allow snake to detect and ambush passing prey, while remaining buried beneath soft sand to avoid

detection. Extends north to Egypt, but locally restricted to semi-arid regions in northern and eastern Kenya, with an isolated population around Dodoma. Juveniles feed mainly on lizards, adults on small mammals and birds. Gives birth to up to 20 young. Usually gentle, but some individuals bite readily. Can emit a foul-smelling anal fluid.

ADDERS AND VIPERS Family Viperidae

Venomous snakes with *hinged front fangs* that become erect as mouth opens. In other ways, however, are relatively primitive and are unrelated to other venomous snakes. *Head covered in small scales* (except in primitive night adders). Most give birth to live young. Bites of many species may lead to serious medical emergencies, often resulting in death.

RHOMBIC NIGHT ADDER
Causus rhombeatus 40–80cm

A stout snake, although thin for an adder, with *rounded snout* and *soft, feebly-keeled scales.* Has *large paired scales on top of head.* Brown-pink body and tail have 20–30 pale-edged dark rhombic blotches. Has

characteristic *dark 'V'-shape* on back of head. Inhabits moist areas in western regions, emerging at night to feed on toads. Range extends southwards to South Africa. Lays 15–26 eggs at start of wet season. Threatens readily, and mild venom causes swelling and pain, but few if any deaths have been recorded.

SNOUTED NIGHT ADDER
Causus defilippii 30–40cm

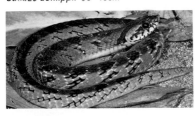

Similar in build and coloration to previous species, although smaller and with pattern of *rounded, triangular dorsal blotches.* Snout also *pointed and upturned.* Prefers

moist savanna, coastal thicket and lowland forest edge. Locally restricted to southern and coastal regions of Tanzania, extending south to northern South Africa. Feeds almost exclusively on small amphibians. Lays small clutches of up to nine eggs. Venom mild, and the few bites recorded caused local swelling and pain.

GREEN NIGHT ADDER
Causus resimus 40–75cm

Chris Kelly

A small, stout adder, typically velvety green with vague darker markings. *Blue skin between scales* revealed when snake inflates body in threat. *Snout slightly upturned.* Usually terrestrial, but may climb into overhanging sedges. Found in scattered spots in the moister western regions, with isolated population in coastal Kenya. Feeds exclusively on frogs and toads. Lays up to 12 eggs at regular intervals during wet season. Venom probably mild, but no case histories known.

FOREST NIGHT ADDER
Causus lichtensteinii 35–60cm

A small, fairly thin, green (but sometimes dark olive to brown) adder with vague dark-centred rhombic markings (may be reduced to chevrons). Has a *white 'V'-shape on back of head.* Snout *slightly upturned* and has large paired scales on top of head. Found in scattered spots in moister regions of Uganda. A terrestrial species that shelters among leaf litter on forest floor. Feeds exclusively on frogs and toads, and lays up to eight eggs. Venom probably mild, but no case histories are known.

NORTHEAST AFRICAN SAW-SCALED VIPER
Echis pyramidum 40–70cm

Steve Spawls

Small, irascible adder that is *boldly blotched*, with small *triangular head* and distinct neck. *Lateral body scales oblique with serrated keels.* Rubs these together to give *rasping hiss* as threat display. These vipers restricted to northern arid regions of Kenya, extending north to Egypt. Unusually for snakes, they eat mainly invertebrates, such as scorpions. Lay up to 20 eggs. Bite readily, and venom causes severe bleeding, frequently leading to death.

KENYA HORNED ADDER
Bitis worthingtoni 30–50cm

Colin Tilbury

A small, stout adder, with a thin neck and *broad, triangular head covered in small scales. Prominent horn occurs above each eye*, and has dark arrow-shape on top of the head. Body scales strongly keeled. *Scales beneath tail unpaired.* Grey body has thin, pale, lateral stripe bordered above and below with dark geometric blotches. Restricted to high grassland in Gregory Rift Valley, Kenya, where it favours rocky, shrub-covered ground and is mainly active at dusk. About 7–12 young born Mar–Apr at the start of rains. Strikes readily when disturbed. Bite appears mild, but few case histories known.

PUFF ADDER

Bitis arietans 80–120cm (up to 185cm in northern parts)

Top: *Subadult (Dodoma, Tanzania)*
Above: *Adult (northern Mozambique)*

A large, thick-bodied, sluggish snake with a *short tail* and *flat, triangular head* covered in *small scales*. Body yellow-brown to light brown, with numerous *dark pale-edged chevrons* on back. Males more brightly coloured than females. Hides at base of a bush to ambush small mammals. Found throughout region, but absent from extreme desert and high mountains. Usually 30–40 young born in rainy season, but an exceptional Kenyan female gave birth to 156 young, the greatest known number of offspring produced by any snake, including anacondas and giant pythons. Very dangerous and responsible for many bites, although with good management these are rarely fatal. Gives deep warning hiss (inaccurately called a 'puff').

GABOON VIPER
Bitis gabonica 80–120cm

A large, heavy adder with a *triangular head* covered in *small scales*, and with *pair of small, horn-like scales* on snout. *Body has attractive geometric pattern* of purple, brown and other pastel colours, while pale head has *thin dark central line.*

This imparts perfect camouflage when the snake shelters among leaf litter on the forest floor. Has patchy distribution in evergreen forests of coastal Tanzania and western Uganda. Up to 43 young, measuring 25–30cm, born in wet season. Docile and rarely bites, but bites should be considered extreme medical emergencies.

NOSE-HORNED VIPER
Bitis nasicornis 60–100cm

A large adder, although not as thick-bodied as Gaboon Viper. *Narrow, triangular head* covered in *small heavily keeled scales,* and nose bears cluster of *2–3 pairs of small horn-like scales.* Body has stunning *geometric pattern* with dark crimson triangles on flanks, and yellow-bordered pale blue rhombic shapes along backbone. Head bears *narrow, dark arrow mark.* Despite almost garish coloration, is perfectly camouflaged in leaf litter on forest floor. Restricted to forest patches of extreme western Uganda and northeast Tanzania. Up to 38 young are born, measuring 18–25cm in length. Irascible, bites readily, and also huffs and puffs in threat.

HORNED BUSH VIPER
Atheris ceratophora 35–55cm

Colin Tilbury

Small, slender bush adder, with a *triangular head* and *prominent horns above each eye*. Coloration very varied, but commonly yellow with irregular black bars and blotches. Some adults uniform light grey to olive, but bright yellow-green when freshly shed. Juveniles black with bright yellow tail tip. Arboreal and often found coiled in vegetation up to 2m above ground. Restricted to forest and forest edges at low and medium altitudes in Eastern Arc Mountains (Usambara, Udzungwa and possibly Uluguru) of Tanzania. A few young born Mar–Apr. The few bites known have been mild, with no fatalities.

MOUNT KENYA BUSH VIPER
Atheris desaixi 40–70cm

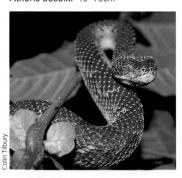

Colin Tilbury

Large, heavy-bodied bush adder with a *broad, triangular-shaped head* and *smallish eyes*. Body green-black and most scales yellow-edged, giving speckled appearance. Some specimens also have lateral yellow crescent-shaped marks that may fuse to form zigzag pattern on hindbody. Juveniles more brightly coloured, with yellow to yellow-green bodies and white tail tip. Arboreal and slow-moving. Juveniles probably feed on amphibians and lizards, while adults take small mammals. Restricted to two isolated evergreen forest patches in foothills of Mt Kenya. Up to 13 young born in Aug. Bites have caused considerable pain and swelling, but not death.

ROUGH-SCALED BUSH VIPER
Atheris hispida 40–60cm

Mark-Oliver Rödel

A slender, small-headed, large eyed bush viper with *long, soft, heavily keeled scales on head and forebody* that give an almost shaggy look. Vaguely blotched body usually greenish-brown in males and yellowish or olive brown in females. An expert climber – favours tall vegetation, such as reeds or papyrus, and creepers in which to bask. Feeds mainly at dusk, moving into lower vegetation from which it ambushes frogs. Restricted to forests in Ugandan region. Bears small litters of 2–12 young. Venom probably relatively mild, but no bites have been recorded.

GREEN BUSH VIPER
Atheris squamigera 40–60cm, exceptionally 80cm

A large, slender bush adder with very variable coloration. Body usually *green to yellow-green, becoming bluer towards tail* with fine yellow crossbars. Belly greenish-blue with yellow blotches. Usually found basking in thick vegetation, up to 6m above ground. Descends to low bush at dusk, ambushing small mammals and frogs from above. Restricted to scattered forests in northwest and westwards to Ghana. As many as 7–9 young born in Aug. Venom causes pain, swelling and incoagulable blood. Deaths are known, but rare.

GREAT LAKES BUSH VIPER
Atheris nitschei 45–75cm

A large bush viper that has a *large, triangular head* and *longish, strongly prehensile tail. Stout green body bears numerous black blotches* that may form irregular bars. *Flat top to head usually has dark arrow-shape* and a dark line behind eye. Basks in vegetation, but hunts lizards and small mammals from low bushes. Found in Albertine Rift evergreen forests above 1,000m, often in papyrus swamps and along waterways. Up to 13 young born Oct–Feb. Often bad-tempered and strikes readily. Venom and bites unknown, but snake should be treated with care.

MATILDA'S BUSH VIPER
Atheris matildaei 40–64cm

Medium-sized bush viper with a *large triangular head* and *longish, strongly prehensile tail. Elongate dark body* bears *yellow rhombic pattern* on back and is yellower on flanks and sides of head.

Flat, dark-topped head has *2–3 prominent horns above each eye*. Basks in vegetation, but hunts lizards and small mammals from low bushes. Restricted to remote montane forest fragments in southern highlands of Tanzania. Considered Critically Endangered, and little known of its biology. Venom and bites unknown.

BARBOUR'S BUSH VIPER
Atheris barbouri 30–37cm

Luke Mahler

An unusual, small, dull-coloured bush viper, with a *triangular head* and *short, round snout. Tail short* and *non-prehensile. Body squat,* dull olive brown with faint pale brown reticulate pattern. Little known of its biology, and remains uncertain whether it lays eggs or has babies. Most specimens have been found along forest edges or clearings. Previously thought to be restricted to Udzungwa and Ukinga ranges of Tanzania, but recently collected in southern highlands.

KENYA MONTANE VIPER
Montatheris hindii 30–35cm

Small, slender viper with a *narrow head covered in small scales.* Males have grey body (brown in females) with series of *paired, pale-edged, triangular patches* on either side of backbone. Grey-white

belly has dark speckles. Lives in grass tussocks in montane grasslands of the Aberdare Mountains and Mt Kenya. Small lizards form main diet. Gives birth to 2–3 small young. Relatively shy and secretive. Bites have been mild, but nothing known of its venom. Endemic to Kenya and currently of conservation concern.

COBRAS AND MAMBAS   Family Elapidae

Fairly large snakes with large *hollow front fangs* that are *not hinged*. Scales mostly smooth; head has large symmetrical scales but *lacks loreal scale* behind nostril. All East African species, except the sea snake, lay eggs. Many have paralysing venom, but spitting cobra venoms also causes necrosis.

BOULENGER'S GARTER SNAKE
Elapsoidea boulengeri 50–76cm

Left: *Adult* Right: *Juvenile*

Large, stout, with a short tail and no obvious neck. *Rounded snout* and *front fangs. Body scales smooth and shiny*, in only *13 rows at midbody*. Juveniles have white head and black back with 12–17 narrow white bands that darken in centre, leaving very narrow paired bands that eventually disappear. Large adults *uniform blue-black with cream bellies*. Shy. Eats small vertebrates, including snakes. Just enters southeastern Tanzania, with scattered records in west. Lays up to eight eggs Jan–Feb. Hisses but doesn't rear or spread hood. Bites cause nasal congestion, headache and muscle pain, but no fatalities known.

EAST AFRICAN GARTER SNAKE
Elapsoidea loveridgei 40–65cm

Similar to previous species in appearance and replaces it in north. Juveniles *usually with pale-edged, pinkish bands* on blue-black body. With growth, bands usually darken in centre, but pale edges usually persist as *paired narrow white bands* (but bright salmon in some northern Tanzanian adults). Lays a few elongate eggs. Biology and toxicity presumably similar to those of Boulenger's garter snake.

USAMBARA GARTER SNAKE
Elapsoidea nigra 40–60cm

Colin Tilbury

Boldly marked garter snake with pale-edged dark bands, which in juveniles are separated by *wider grey bands*. These darken with age, so adults are dark with thin, paired, white bands. Tail short (5–7% body length). A slow burrower in leaf litter, seems to specialise in eating caecilians (p.153). Inhabits evergreen mountain forest. Restricted to the northern parts of the Eastern Arc Mountains of Tanzania and the Shimba Hills, southern Kenya. Lays small clutch of up to five eggs Oct–Jan.

YELLOW-BELLIED SEA SNAKE
Hydrophis platurus 60–80cm

Sea snakes common in East Indies, but only this species enters African coastal waters. Inhabits Indo-Pacific region, swimming in waters of the Mozambique Channel between East Africa and Madagascar and usually seen only when washed onto beach by strong onshore winds. *Bright yellow and black stripes, flat head and oar-like tail* unmistakable. These snakes drift in surface currents, ambushing small fish sheltering in floating debris. From 3–5 young are born in surface waves. Powerful venom causes paralysis, but only a single death recorded.

ASHE'S SPITTING COBRA
Naja ashei 180–280cm

Wolfgang Wuster

Steve Spawls

A very large, thick-bodied cobra that spreads a *broad hood* and *spits venom*. Usually has *over 195 ventrals* and *21+ scale rows on neck* but *no subocular scale*. Adults generally *plain olive brown* with *pale brown belly*, usually with *wide, dark brown band on neck* that *encroaches on sides*. Juveniles light brown, often with faint 'herringbone' pattern, top and upper sides of head and neck greyish-brown with dark, more prominent throat band. Inhabits dry lowland regions of northern and coastal Kenya, just entering northern Tanzania. Terrestrial and usually forages in early evening, but juveniles sometimes active by day. Biology poorly known. One very large female killed and swallowed a domestic cat! Produces prodigious quantities of venom, but case histories unknown, although venom probably similar to that of Black-necked Spitting Cobra.

BLACK-NECKED SPITTING COBRA
Naja nigricollis 130–220cm ☠

Mark-Oliver Rödel

A Large, thick-bodied, nocturnal cobra that spreads a *broad hood* and *spits venom*. Has *no subocular scale. Coloration varied.* Adults usually plain olive brown to black (sometimes red-brown or copper-coloured), with dirty yellow or grey to black belly that may also be heavily speckled. Juveniles grey-brown with black head and neck. Inhabits savanna to semi-arid scrub, never thick forest. Widespread in northern parts, but rare in southern Tanzania. Mainly terrestrial and usually forages in early evening (juveniles sometimes active by day). Eats most small vertebrates and lays up to 20 eggs. Venom causes swelling, pain and local necrosis, as well as blood-clotting abnormalities, and fatalities have been known.

RED SPITTING COBRA
Naja pallida 80–130cm

Left: *Adult* Right: *Juvenile*

Small, relatively slim, brightly coloured cobra that *spreads a narrow hood* and *spits venom*. Has *no subocular scale*. Juveniles *bright red-brown* with distinct *black throat band*. Colour in adult darkens to dull red-brown and throat band may disappear. Found in dry savanna and semi-arid desert in northeast of region. Terrestrial and mainly nocturnal, hunting wide range of small vertebrates, but especially frogs. Lays small clutch of 6–15 largish eggs. Venom causes pain, swelling and necrosis but rarely death.

MOZAMBIQUE SPITTING COBRA
Naja mossambica 90–130cm

Small, short-tempered spitting cobra with a blunt head and *23–25 midbody scale rows*. Pink-grey to dark olive body has black-edged scales, while pinkish belly may have irregular black crossbands or blotches on throat. Diet varied and includes mice, lizards and amphibians. Even grasshoppers may be eaten. Restricted to southeastern Tanzania and south to Zululand and Namibia. Lays 10–22 eggs in rainy season. Due to its nocturnal habits, people do not see it in the dark, resulting in many snakebites. Fortunately fatalities are rare. When disturbed, spreads a broad hood and spits readily.

BANDED WATER COBRA
Naja annulata 150–270cm

Large, stout, aquatic cobra of the Congo Basin from Cameroon to Angola. In the east restricted to Lake Tanganyika. Juveniles golden brown with paired, narrow, dark bands that are retained in adults in Congo Basin. In specimens from southern Lake Tanganyika, where it is now very rare, bands fade with age, and adults become plain brown, darker towards tail, with 2–7 thin, dark stripes on neck. A powerful swimmer, feeding mainly on fish. Dives easily to depths of 30m, and may stay submerged for up to 10 minutes. When not foraging, shelters in shoreline rocks, or even under jetties and partially submerged boats. Shy, retreats quickly, and bites very rarely. Nothing known of its reproduction or venom.

FOREST COBRA
Naja melanoleuca complex 180–280cm

Eastern savanna phase

The largest cobra in Africa. Heavy but slender body has *19 rows of glossy scales*. Head and forebody yellow-brown, heavily speckled with black. Shiny blue-black tail. *Lip scales mostly white with black edges*. Usually has 2–3 black throat bands, but some populations lose bands and belly develops heavy speckling. Found in evergreen forest or moist savanna. Occurs patchily in coastal forests of Tanzania and southern Kenya, and in forest patches north of Lake Victoria (probably different species). Mainly diurnal, hunts small vertebrates on forest floor. May climb trees (up to 10m), is also fond of water and will eat fish. Lays 15–26 large eggs. When cornered, rears and spreads narrow hood. Bites rare but serious and may be fatal.

EYGPTIAN COBRA
Naja haje 130–250cm

A large cobra that spreads a broad hood, but *does not spit. Large subocular scale separates eye from upper labials.* Juveniles usually dirty golden-brown (sometimes reddish or yellowish) with many fine irregular bands above, golden bands below, and prominent black throat band. Adults grow darker, with paler belly and faded throat band. Restricted to scattered localities in central and northwest regions. Terrestrial. Hunts mainly at twilight, but may be active by day or at night. Juveniles eat toads, lizards and small mammals. Adults often also take bird eggs. Very large cobras eat mammals and snakes, particularly puff adders. Lays up to 20 large eggs. Venom causes paralysis and often death.

GOLDIE'S TREE COBRA
Pseudohaje goldii 150–270cm

Very large, arboreal, with a *short, blunt head and very large eyes.* Spreads very *narrow hood,* but does not spit. Has large, glossy body scales in only *15 rows at midbody.* Top of head and back black. Belly yellow extending onto lower flanks. Sides of head and neck yellow, with black scale margins. Juveniles may have a few irregular yellow crossbands on forebody. Active in forest by day, perhaps also at night, hunting small vertebrates and possibly fish. Occurs throughout Congo forest, just entering Uganda. Lays 10–20 eggs in a rotting tree or other suitable spot. Venom reportedly very toxic, but no case histories known.

GREEN MAMBA
Dendroaspis angusticeps 180–230cm

Smaller, more slender and *arboreal* cousin of Black Mamba. Body *brilliant green* and *mouth lining pink-white*. Patchily distributed throughout coastal forest from Witu south, and inland in Tanzania throughout forest patches. Lives in upper forest canopy, searching for small mammals and birds that it catches by ambush. Lays up to 10 eggs in hollow log or leaf litter. Shy and rarely seen, so bites are fortunately rare and not as dangerous as those of Black Mamba, although they are still potentially fatal.

JAMESON'S MAMBA
Dendroaspis jamesoni 200–260cm

Warren Klein

A large green mamba that replaces the former species in rainforest patches around Lake Victoria. Body *dull green*, with paler belly and *yellow throat and neck*. Head scales large and have *narrow black border*. Lives in forest and woodland, searching upper canopy for small mammals and birds that it catches by ambush. Shy and rarely seen, although sometimes common around villages. Little known of its biology or venom, and bites are rare. Should be treated with care.

BLACK MAMBA
Dendroaspis polylepis 220–320cm

 Africa's most feared snake. A *very large,* slender species with a coffin-shaped head and smooth, but dull, scales. Body dirty grey, sometimes olive, with black blotches on pale grey-green belly. *Mouth lining black.* Found in moist and dry savanna, mainly in eastern and central regions. Active, mainly terrestrial species that occasionally climbs trees in search of food. Rats, hyraxes and squirrels form main diet, although birds also eaten. These are pursued and repeatedly stabbed until they succumb to the very toxic venom. Lays small clutch of 12–18 large eggs. When disturbed, rears forebody, gapes widely, spreads very narrow hood, and gives hollow hiss – a warning best heeded!

AFRICAN SNAKES
Family Lamprophiidae   Subfamily Lamprophiinae

Restricted to Africa, these are characteristically nocturnal snakes that lack fangs and kill by constriction. They are harmless, usually terrestrial, and lay eggs.

RED-AND-BLACK STRIPED SNAKE
Bothropthalmus lineatus 50–100cm

Left: *Adult* Right: *Juvenile*

A brilliantly coloured snake, with *vivid red and black stripes* extending down body and long tail. Head white in juveniles, fading to reddish with age and with unusual *grooves between nostrils and small eyes* to allow for good forward vision. *Body scales keeled.* A forest species, just entering forest patches in the extreme northwest of region. Forages at night on forest floor, feeding on small mammals. Clutches of up to 29 eggs have been recorded. Harmless, but bites readily.

FOREST HOUSE SNAKE
Boaedon olivaceus 45–90cm

A medium-sized, solid snake with head only slightly distinct from body. *Body scales smooth,* in *25–31 rows at midbody,* and *subcaudals*

undivided. Head and body *uniform olive brown,* sometimes blackish. Belly yellowish. *Eyes orange-red.* A forest species restricted to extreme western regions. Nocturnal and terrestrial, ambushing rodents and other small mammals. Like other house snakes, lays eggs.

Marius Burger

BROWN HOUSE SNAKE COMPLEX
Boaedon fuliginosus 70–120cm

Top: *Large-eyed form (Dodoma, Tanzania)*
Above: *Striped form (Arusha, Tanzania)*

Large house snakes, considered to be a single species for many years, but now known to comprise numerous new species. Brown House Snake occurs in southern and eastern parts of region and has *rust-red body*, off-white belly and *pair of thin, yellow-white stripes on sides of head and neck*. Some populations also have large eyes. Individuals from northern Tanzania have stripes that continue along body almost to tail. Sooty House Snake occurs in western and central regions and is dark olive green in colour (almost black in large specimens) with head stripes faint or even absent. All are terrestrial and nocturnal and feed mainly on rodents, such as mice and rats, although juveniles also eat lizards. Tolerant of urban squalor, occurring commonly around houses (hence common name). Up to 18 eggs laid in rainy season. Harmless and kill prey by constriction.

FOREST NIGHT SNAKE
Hormonotus modestus 50–85cm

Shy, secretive and gentle snake of the forest floor. Has a *pear-shaped head* with large eyes that have *vertical pupils*. Body yellow-grey to brownish, smooth-scaled, with longish tail. *Head scales white-edged* with dark spot on each lip scale. Restricted to forest patches in northwest of region, extending through Congo rainforest to Guinea. Little known of its biology, but probably lays eggs and eats small vertebrates such as lizards.

RED-SNOUTED WOLF SNAKE
Lycophidion uzungwense 20–60cm

Wolfgang Böhme

Plain, dark grey snake with a *flat head, small eyes and distinctive orange-red edge to snout*. Body cylindrical and tail relatively short. Has smooth, white-tipped, dark body scales and dark grey belly. Terrestrial and nocturnal. Restricted to the Udzungwa Mountains in central Tanzania. Sleeping skinks and other lizards form main diet. Lays a small clutch of eggs.

JACKSON'S WOLF SNAKE
Lycophidion capense jacksoni 40–58cm

Wolf snakes, which are peculiar to Africa, are named for their long *recurved teeth*, used to prise sleeping prey from their retreats. Nonetheless, a harmless, gentle snake that never attempts to bite. Small with *black body and white scale tips*. Head flat with dark crown and *pale pink snout band*. Eyes small with vertical pupils, and *first upper labial touches postnasal*. Elongate body has 170–211 ventrals. Widely distributed, usually in savanna and farm bush. Diet consists mainly of skinks and sand lizards. Lays 3–8 eggs during wet season.

FLAT-SNOUTED WOLF SNAKE
Lycophidion depressirostre 30–48cm

A small wolf snake with a *very flat head*. Head and *body scales heavily speckled in white*. Belly dark grey, often pale on edges. Found in dry and moist savanna in eastern parts of region. Biology poorly known but probably similar to that of Jackson's Wolf Snake. Harmless, like all wolf snakes, and never attempts to bite.

CAPE FILE SNAKE
Gonionotophis capensis 100–150cm

Unusual and rarely seen snake, easily recognised by thickset, *triangular body* and *very flat head*. Almost conical, *strongly keeled scales* separated by *bare pink-purple skin. Scales along backbone white, enlarged and have two keels.* Body grey-brown with ivory-cream belly and flanks. Locally restricted to coastal and southeastern Tanzania. A formidable predator of other snakes and kills by constriction. Will even tackle venomous snakes, including cobras, to whose venom it is partially immune. Lays clutch of 5–13 relatively large eggs in leaf litter. Docile in disposition and never bites, but may void bowels when handled.

BLACK FILE SNAKE
Gonionotophis nyassae 40–60cm

Flattened head, triangular body and pinkish skin between body scales confirm this is a file snake. Can be distinguished from previous species by uniform purple-black colour, although may have creamish-white belly. In addition, is smaller, has slightly longer tail (51–79 subcaudals) and fewer (less than 184) ventrals. Inhabits savanna and coastal thicket south of Tana River. Although small snakes are eaten, lizards, particularly skinks, form main diet. Up to six eggs are laid. Small, shy and rarely bites. Usually moves jerkily, trying to hide head under body coils.

PLAIN FILE SNAKE
Gonionotophis chanleri 100–142cm

Large file snake with a *triangular body, flat head* and *strongly keeled* body scales, separated by bare dark skin, in 15 rows at midbody. Scales along backbone *black, enlarged*, and *four-keeled*.

Tail relatively short, 12–17% of body length, and has only 44–62 paired subcaudals. Body uniform *black-brown* with *dark belly*. Inhabits dry savanna, from eastern DRC, through northwest Tanzania and Kenya to Eritrea. Inhabits burrows, emerging at night to feed on snakes and sleeping lizards. Nothing known of its breeding.

FOREST FILE SNAKE
Gonionotophis poensis 60–120cm

Typical file snake with a *triangular body, flat head and strongly keeled body scales*. Scales along backbone *black, enlarged* and have *two keels*. *Tail long*, 20–25% of body length, with 75–124 paired subcaudals. Body uniform grey-black to olive brown and lighter below. Inhabits moist evergreen forest, entering extreme western region and extending through Congo forest to West Africa. May be partially arboreal and feeds on lizards and possibly snakes. Nothing known of its breeding.

Subfamily Pseudoxyrhopiinae

Also known as Malagasy snakes. Only three genera occur in Africa.

SLUG EATERS
Duberria lutrix complex 30–40cm

Stout-bodied little snake with *small head* and short tail. Back is brick-red to pale brown, sometimes with broken black line along backbone. Paler flanks, from grey to light brown. Belly cream, edged with dark dotted line. Isolated

Central Serengeti

populations in mountain grassland and moist savanna. Feeds entirely on slugs and snails. Shy, usually hiding in damp situations. Six to nine young born in rainy season. When handled, releases an unpleasant cloacal fluid. Currently treated as a single species.

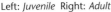

Subfamily Pseudaspidiinae

An unusual subfamily containing only two African species.

MOLE SNAKE
Pseudaspis cana 100–180cm

Left: *Juvenile* Right: *Adult*

A large snake known by its thick body, *slightly hooked snout*, small eyes and round pupils. Body scales smooth, in *25–31 rows at midbody*. Colour varies by region and age. Young light brown with four rows of dark pale-edged spots that often fade in subadults (about 1m), but may persist in some adults. Normally adults light to red-brown,

sometimes grey-olive to dark brown. Sporadic isolated populations in grassland and savanna. Kills rodents and moles by constriction. Up to 95 young born in rainy season. Bite painful, but not poisonous.

Subfamily Prosymniinae

A small subfamily restricted to Africa, with one genus of 16 species.

EAST AFRICAN SHOVEL-SNOUT
Prosymna stuhlmanni 24–28cm

Small, with a *rounded snout* and two postocular scales. *Tail short* with only 17–39 subcaudals. Dorsal scales have *single apical pit* and are in *17 midbody rows*. Dark brown to metallic blue-black body usually has *pale-centred scales*, and *paired small white spots* may flank backbone. Although usually white, belly may be brown-black. Found in wooded savanna in southern regions. Like all shovel-snouts, eats only reptile eggs. Lays 3–4 elongate eggs in leaf litter in rainy season. Shy and secretive and never bites or wriggles violently.

MIOMBO SHOVEL-SNOUT
Prosymna ambigua 8–35cm

A stout shovel-snout with a prominent, angular, *horizontal 'shovel'* on tip of the snout. Body *uniform lead grey*, with each body scale slightly *pale-centred*. Throat and belly paler and each scale, particularly on and towards tail, has pale edge. Biology poorly known, but probably similar to that of other shovel-snouts. Found from southern Sudan to Zambia, and west to Gabon and northern Angola. Prefers miombo woodland and forest edge.

Subfamily Psammophiidae

Also known as sand and whip snakes, members of this subfamily occur mainly in Africa, with a few species in Eurasia and Madagascar. Diurnal, mainly terrestrial, with enlarged back fangs, they kill by injecting venom. Only a few species can harm humans.

RED-SPOTTED BEAKED SNAKE
Rhamphiophis rubropunctatus 140–250cm

Very large, stout-bodied, with a long tail, *smooth* scales in *19 rows at midbody, 130–160 paired subcaudals* and *prominent hooked snout.* Orange on head may extend onto grey to grey-brown body. Juveniles cream with grey dorsal stripe and scattered red spots. Occurs in coastal thicket and scrublands in eastern Tanzania-Kenya border region and in Tana River region, with scattered records further north. A terrestrial diurnal hunter of lizards and small mammals. Presumed to lay eggs and to be harmless.

RUFOUS BEAKED SNAKE
Rhamphiophis rostratus 120–140cm

Large, with a *prominent hooked snout.* Tail relatively *short,* with only *90–125 paired subcaudals,* and heavy muscular body has *17 midbody scale rows.* Colour uniform warm yellowish to red-brown, sometimes with pale-centred scales. Head has distinctive *dark brown eye stripe.* Occurs throughout central and eastern regions, sheltering in burrows in sandy habitats. Eats various small vertebrates, including other snakes. Lays 8–17 large eggs, often staggered over several days. May hiss and strike when first caught, but tames well. Venom not dangerous.

SHARP-NOSED SKAAPSTEKER
Psammophylax acutus 60–100cm

Small, with a *sharply pointed snout* and *large eyes*. Tail relatively *short*, with only *53–72 subcaudals*. Body *boldly striped*. Dorsal stripes extend forward onto head in a *trident shape* (often missing in northern specimens). Cream belly. Inhabits moist grasslands and flood plains where it hunts small mammals and frogs. Isolated populations on north and south borders of Lake Tanganyika and in extreme western Uganda. Lays 10–15 elongate eggs.

KENYAN STRIPED SKAAPSTEKER
Psammophylax multisquamis 60–120cm

A *boldly striped*, medium-sized snake with a *short, somewhat pointed head*, 17 scale rows at midbody and *relatively short tail*. Cream to golden tan
with bold, wide, dark stripe along flanks and sides of head, and thinner, light brown stripe along backbone. Restricted to Kenyan high grasslands, with scattered records elsewhere. Active by day, hunting small vertebrates. Lays up to 16 eggs, which female protects by coiling around them.

STRIPED SKAAPSTEKER
Psammophylax tritaeniatus 60–80cm

Slightly smaller, but similar in build to previous species, an attractive snake distinguished by its more *pointed snout*. Also has *boldly striped pattern*, with three black-edged dark brown stripes on pale grey body. Middle stripe may be divided by fine yellow line. Upper lip and belly plain white. Hunts small mice in southern savanna of Tanzania, but will also take frogs and lizards. Lays up to 18 eggs in a hole in rainy season, but female does not guard them.

DWARF SAND SNAKE
Psammophis angolensis 30–40cm

A *tiny, elegant,* rarely seen sand snake. Has only *11 midbody scale rows* and *conspicuously striped body*, with broad brown-black dorsal stripe and faint broken black stripe on lower flank. Dark brown head has three *narrow crossbars* and one or several dark neck collars. Found from South Africa to southern savannas of East Africa. Forages among grass tussocks and fallen logs in lowveld savanna, where it feeds on small lizards and frogs. A gentle, shy species that seldom bites, but unfortunately does not settle well into captivity. Lays small clutch of 3–5 elongate eggs in moist soil under a log.

LAKE RUKWA SAND SNAKE
Psammophis rukwae 80–120cm

Colin Tilbury

A large, slender sand snake with *17 midbody scale rows* and *boldly striped* pattern. Belly *yellow, with thin black lines* on sides. Easily confused with Eastern and Northern stripe-bellied sand snakes, but has five (not four) *lower labials* touching chin shields. Another active hunter of lizards in the central savannas of Tanzania and adjacent regions. Details of venom and reproduction poorly known, but is almost certainly egg-laying. Taxonomy remains controversial.

EASTERN STRIPE-BELLIED SAND SNAKE
Psammophis orientalis 90–120cm

A *thin, medium-sized* snake with a *long head, large eyes* and *very long thin tail.* Usually obviously striped in southern Tanzania, but uniformly coloured further north. Has distinctive *yellow belly, bordered by black and white stripes.* Often seen moving head up, alert for prey, but may also climb into low bushes. A fast, active hunter of birds, lizards and mice in coastal grasslands and mesic thicket. Difficult to catch due to its speed and mobility, although many are caught by birds of prey. Lays about 4–10 elongate eggs in an underground tunnel.

NORTHERN STRIPE-BELLIED SAND SNAKE
Psammophis sudanensis 70–120cm

Similar to previous species, but distinguished by *crossbars on back of head*. Body boldly striped in brown and gold, and has *yellow belly* bordered by black stripe. Smooth body scales in *17 rows at midbody*. An active diurnal hunter of lizards; mainly terrestrial, but moves easily in low bushes. Lays 4–10 eggs Jan–Feb.

SPECKLED SAND SNAKE
Psammophis punctulatus 100–180cm

Another long, thin, *boldly striped* sand snake, but with *heavily speckled lower flanks*. Tail long (33% of body length) and thin. Has smooth

body scales in *17 midbody rows*. Back yellow (grey in juveniles) bearing three black stripes, that along backbone being widest. *Narrow head, orange* above with *white lips*. Very active, fast-moving hunter of small lizards. Found in dry and semi-arid savanna in northern Kenya, just reaching Tanzania. Lays up to 12 eggs Jul–Aug. Bites readily, causing local pain and swelling.

LINK-MARKED SAND SNAKE
Psammophis biseriatus 50–100cm

Colin Tilbury

Chris Kelly

A slim, grey sand snake with *olive brown dorsal stripe edged with black dashes*. Only two *upper labials* enter eye. Tail very thin and nearly 40% of body length. Smooth body scales in *15 midbody rows*. Head narrow and brown above, with dark line through eye. Has white lips and orange-brown temporal region. Active during the day, chasing lizards, its main prey. Lays small clutches of 2–4 elongate eggs Jul–Sept. Bites may cause local pain and swelling.

TANGANYIKA SAND SNAKE
Psammophis tanganicus 60–100cm

Chris Kelly

Once considered a southern race of Link-marked Sand Snake, this *thin* snake differs in having three *upper labials* entering eye. Top of head boldly patterned with *dark-edged tan blotches on paler background*. Body has broad grey to brown vertebral band, edged with dark blotches that may form crossbars. Has smooth body scales in *15 midbody rows*. Very long, thin tail helps it climb in low scrub. A very active hunter of lizards in the central savanna of Tanzania. May shelter in low bush or under rock cracks. Little known of its biology or venom.

OLIVE GRASS SNAKE
Psammophis mossambicus 100–140cm

Steve Spawls

A *big, robust* snake, with a long tail and large scales in *17 midbody rows*. Usually has *more than 164 ventrals*. Olive brown back grows paler towards tail. May have scattered black flecks on sides of forebody. Body scale sometimes black-edged to form thin black stripes. White-yellow belly may have black streaks. Widely distributed, but absent from northeastern Kenya. An active diurnal hunter that often moves with forebody lifted. Eats various small vertebrates, including other snakes. Lays up to 30 eggs in a hole in ground. Has nervous disposition, retreating swiftly into cover, but will bite readily when caught. Mild venom may cause nausea, but is not dangerous.

BARK SNAKE
Hemirhaggerrhis nototaenia 25–40cm

A pretty, secretive snake. Shelters under loose bark or in hollow trees in southern savanna regions. Small and slender, with a distinct *flattened head* and large eyes with *vertical pupils*. Back grey with *dark dorsal stripe* flanked by, and sometimes fused with, series of black spots. Small lizards, particularly dwarf day geckos (p.113), form main diet, although small tree frogs may also be eaten. Food swallowed as snake hangs head-down from branches. Lays small clutch of 2–8 elongate eggs in a tree hollow. Harmless.

STRIPED BARK SNAKE
Hemirhaggerrhis kelleri 30–42cm

A small, docile and harmless snake with a *flattened head, sharp snout* and *small eyes with large vertical pupils*. Body plain light grey to brown with *broad, dark dorsal stripe* that extends onto tail. Dark lateral stripe may also be present. Head has

dark lateral stripe running through each eye to jaw. Belly pale with *three dashed stripes*. Shelters in hollow trees or under debris and feeds on small lizards, particularly dwarf day geckos. Some populations also eat gecko eggs. Inhabits arid savanna from northern Tanzania through Kenya to Sudan, Ethiopia and Somalia.

Subfamily Atractaspididiinae

Members of this subfamily, also known as African burrowing snakes, are peculiarly African. Most are back-fanged (except for one species), while burrowing asps have evolved erectile front fangs and unique venom. They push through loose soil or sand, mainly hunting other burrowing reptiles.

FAWN-HEADED SNAKE EATER
Polemon collaris 50–85cm

A dark-bodied burrowing snake of Congo rainforest. Head not distinct from neck, *snout is rounded* and *eyes are small*. Shiny, *smooth body scales* in *15 midbody*

rows*, and *tail short and stumpy*, ending in a *spike*. Distinguished by *pale fawn to orange band on back of head*. Body iridescent grey-brown above, each scale having a dark edge, with grey-cream belly. Burrows in humic leaf litter, hunting blind snakes, which seem to be its sole diet. Its biology and venom are poorly known.

GERARD'S BURROWING SNAKE
Chilorhinophis gerardi 35–45cm

Colin Tilbury

A very *thin-bodied* burrowing snake, with *head not distinct from neck*. Has a rounded snout and small eyes. Smooth body scales in *15 midbody rows*, and tail short and stumpy, ending in a spike. *Body*

boldly striped in black and pale yellow*, with black neck band and *bright yellow belly*. Tail black like head, and is held upright and waved to mimic the head. Burrows in humic leaf litter, hunting thread snakes, which seem to be main diet. Biology and venom poorly known.

CAPE CENTIPEDE EATER
Aparallactus capensis 25–35cm

Small, slender and rarely seen, although often locally common. Head small with *rounded snout* and *prominent black collar*. Body varies in colour from *red-brown to grey-buff*, with cream belly. Restricted to the southern regions, extending to South Africa. Spends most of its life underground in tunnels, or in rotting logs or rock piles, and particularly in old termite nests, where it searches for centipedes. When seized, these quickly succumb, although the venom is completely harmless to humans. Lays 2–4 small elongate eggs.

BLACK CENTIPEDE EATER
Aparallactus guentheri 30–45cm

Although similar in build and habits to previous species, this snake differs in having a *blue-grey to black* head and

Top: *Adult* Above: *Juvenile*

body with *two narrow yellow collars* on neck. Chin and belly off-white. Hunts scorpions and centipedes under logs and stones in patches of moist evergreen forest along coast and is associated with the Eastern Arc mountains in Tanzania. Known to lay eggs, but further details uncertain.

GIANT CENTIPEDE EATER
Aparallactus modestus 35–65cm

Large, *heavy-bodied* centipede eater with a *short tail*, rounded snout, and no obvious neck. Scales beneath tail are *unpaired*. Adults *uniform blue-green*, sometimes with light-edged scales. Juveniles have pale

head patch that quickly fades with growth. Inhabits forest of Congo Basin, reaching eastern limit in Uganda. Burrows in soft forest soils and is restricted to forest patches in extreme northwest of region. Thought to feed on soft-bodied prey such as earthworms or slugs, and lays small clutch of 4–7 elongate eggs. Harmless.

JACKSON'S CENTIPEDE EATER
Aparallactus jacksonii 22–28cm

A small centipede eater with a *short tail*, rounded snout and no obvious neck. *Scales beneath tail unpaired*. First lower labials are in contact below snout, and *3rd–4th upper labials enter eye*. Adults uniform orange-brown, with broad black neck band, often *bordered with yellow behind*. Inhabits coastal bush and savanna along eastern Kenya–Tanzania border region, with scattered records elsewhere. Burrows in loose soil and feeds on centipedes. Unlike other centipede eaters gives birth to 2–3 young. Harmless.

COMMON PURPLE-GLOSSED SNAKE
Amblyodipsas polylepis 40–68cm

A stocky snake, with a *blunt head and small eyes*. Has *17–19 midbody scale rows* and blunt tail with *subcaudals in 17–29 pairs*. Black with *smooth scales* that have

attractive *purple gloss* after shedding. Found in moist coastal regions or moist savanna and evergreen forest. Other burrowing reptiles form its main diet, particularly blind snakes (p.11). Small prey simply swallowed alive, larger prey subdued by constriction. Lays up to seven large eggs. Gentle and rarely bites.

MPWAPWA PURPLE-GLOSSED SNAKE
Amblyodipsas dimidiata 35–52cm

Chris Kelly

A *thin* and *shiny-scaled* burrowing snake with a *long snout, underslung mouth* and *short tail*. Body scales in *17 rows at midbody*. Typically iridescent purple-brown above, with *bright yellow flanks* and pinkish-yellow belly. Restricted to dry savanna in central Tanzania, where it burrows in loose sand. Biology poorly known. Probably feeds exclusively on worm lizards and lays eggs.

BIBRON'S BURROWING ASP
Atractaspis bibronii 40–70cm

A thin burrowing snake with *smooth, shiny black scales* in *21–23 rows at midbody*. Head has slightly pointed snout, *underslung mouth* and *long, partially erectile front fangs*. Belly may be white or dull black. Tail short with *undivided subcaudals* and a *terminal spine*. Usually found sheltering in rock piles, in old termitaria or under rotting logs, where it searches for other burrowing reptiles. Nestling mice may also be eaten. Lays up to seven elongate eggs. When disturbed, holds head and forebody in a tight, vertical swan-neck shape, trying to push its head into soil. Cannot be held safely, as it can erect a single fang and strike backwards into the hand holding it. Venom causes swelling and extreme pain, but deaths are unknown.

VARIABLE BURROWING ASP
Atractaspis irregularis 30–66cm

As with all burrowing asps, head has a slightly pointed snout, *underslung mouth* and *long, partially erectile front fangs*. Smooth, iridescent body scales in *23–27 midbody rows*. Short tail ends in a *spine* and *subcaudals are paired*.

Body usually glossy black above and grey-black below. Restricted to forest patches in western regions, where it burrows in humic soil, feeding on rodents and probably other burrowing reptiles. Lays up to six elongate eggs. Venom causes swelling and extreme pain, and there have been several recorded fatalities.

SMALL-SCALED BURROWING ASP
Atractaspis microlepidota 50–110cm

A heavy-bodied burrowing snake with *small scales*, a slightly pointed snout, and *long, partially erectile front fangs*. Its body scales are smooth, in *27–35 rows at midbody*, and it is usually a dull purple-black above and below. The short tail has *undivided subcaudals* and a *terminal spine*. Burrows in sandy soil and feeds on other burrowing reptiles and small rodents. Lays up to eight elongate eggs. Bites readily when caught, and the venom causes swelling, extreme pain and often death.

TWO-COLOURED SNAKE
Micrelaps bicoloratus 20–32cm

A tiny, thin burrowing snake with a rounded snout, smooth body scales in *15 midbody rows*, and *boldly striped pattern*. Tanzanian snakes have lateral brown stripes, while Kenyan snakes have broad median dark brown stripes. Belly white to cream, sometimes with black median line beneath tail. Burrows in loose soil, searching for blind snakes and legless skinks, which form main diet. Little known of its biology and venom. Its relationships are unknown; is now not thought to be related to African burrowing snakes.

OLD WORLD SNAKES Family Colubridae

Mostly back-fanged snakes of Eurasia and Africa, just entering the New World. Relationships still not fully determined and family may be further split. Nocturnal and diurnal species inhabit diverse habitats. Some species have venom of clinical importance.

Subfamily Natricinae

Palaearctic water snakes: only three genera occur in Africa.

FOREST MARSH SNAKE
Natriciteres sylvatica 30–40cm

Small, thin, with *smooth scales* and a *relatively long tail* (60–84 subcaudals). Coloration varies. Dark olive to chestnut-brown back may have vague blotches, a darker vertebral band, or faint yellow collar. *Lips usually barred.* Found in forest marshes in southern Tanzania, often in dead logs. Frogs, particularly squeakers (p.134) and puddle frogs (p.146), form main diet. Lays up to six eggs in moist leaf litter in rainy season.

OLIVE MARSH SNAKE
Natriciteres olivacea 30–54cm

A medium-sized marsh snake with *black-edged yellow lip scales,* smooth body scales and a *relatively long tail.* Body may vary from bright green to blue-black, or even light brown, usually with broad *dark band along backbone.* Widespread, but absent from drier central and northern regions. Lives along streams and around ponds. Feeds on frogs and fish, and lays up to 11 eggs.

Subfamily Grayiinae

African water snakes are large snakes of the Congo Basin and West African forest, just entering the western parts of the region.

SMITH'S WATER SNAKE
Grayia smythii 100–200cm

A large water snake. Easily confused with Banded Water Cobra (p.31) due to its *banded body* and *dark-edged head scales*, but its *smooth body scales are in only 17 rows at midbody*, and it lacks front fangs.

Juveniles dark-bodied with lighter bands that divide on lower flanks. With age, pattern reverses, forming a series of dark saddles on back with lighter gaps on flanks. *Spots on yellow belly often in long, thin rows.* Found in aquatic habitats in northwest, around Lake Victoria. A superb swimmer: hunts at night and in low light. Catches fish and frogs in quiet pools and bays. Lays up to 20 large eggs.

THOLLON'S WATER SNAKE
Grayia tholloni 70–110cm

Kenny Babilon

Medium-sized water snake with *body bands* that are *lighter on flanks*. Lip scales are *dark-edged*. Smooth body scales in *15 rows at midbody*. Belly yellow with *black lateral spot* on each scale. Found in Congo Basin, extending into southern Sudan, northern Angola, and Great Lakes region. Hunts fish and frogs in sheltered pools in low light. May be viviparous.

Subfamily Colubrinae

This subfamily includes many of the typical snakes of the northern hemisphere. Well represented in Africa, with many groups lacking fangs, others back-fanged and some even highly venomous.

SMITH'S RACER
Platyceps smithii 40–70cm

Steve Spawls

A small, slim snake with *smooth scales* in *21 midbody rows* and longish tail. Light brown body strongly barred in juveniles, but bars break up into spotted rows with age. Has three prominent black bars across the eyes, temporal region and neck. Belly varies from white to yellow. Diurnal and terrestrial, although it may climb into low scrub in dry savanna and semi-desert. Lizards form main diet. Lays 2–4 elongate eggs. Probably harmless.

SEMIORNATE SNAKE
Meizodon semiornatus 40–60cm

A *small, slender* snake with *irregular black crossbars* on forepart of grey-olive body. Flat black head has *relatively large eyes* partially ringed in white and with *round pupils*. Restricted mainly to eastern Tanzania, with scattered records further north, where it inhabits margins of pans and marshes, feeding on small frogs. During dry season, shelters in hollow logs and under dead tree bark, where several snakes may be found together. Lays only a few large, elongate eggs. Rarely bites and is considered harmless.

CROSS-BARRED TREE SNAKE
Dipsadoboa flavida 60–80cm

 Tree snakes are nocturnal and back-fanged. This small, slender species has a wide, *heavily blotched* head, large eyes 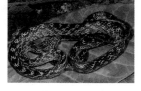 and vertical pupils. Has *17 scale rows at midbody*. Tongue *white-tipped with black bar* at the fork. Body red-brown, heavily patterned with *58–82 light yellowish crossbands*, decreasing in size and number with age. Restricted to the coastal region. Shelters in woodland and palm thicket during the day, emerging at night to feed on geckos and reed frogs. Lays eggs, but biology is poorly known. When first caught, is very willing to bite, adopting an open-coiled posture, but is harmless.

GÜNTHER'S TREE SNAKE
Dipsadoboa unicolor 80–128cm

Wolfgang Böhme

A long, slender snake with a *wide, flat* head and large eyes with vertical pupils. Has *17 scale rows at midbody*. Juveniles and subadults pale brown-grey, but darker towards tail. With age body becomes green with bluish tail, then olive black. Some specimens have white spots on scales of forebody. Restricted to extreme western region, and thence to West Africa. Nocturnal in habits, it frequents upland forests where it feeds on frogs. Lays eggs, but biology poorly known. Bite considered harmless.

WERNER'S TREE SNAKE
Dipsadoboa werneri 70–120cm

A large tree snake with a *broad head* and long, thin body and tail. Eye large with *vertical pupil* and *yellow-green iris.* Has *19 scale rows* at midbody. Body varies from pale grey to red-brown, often with pale-centred scales. A nocturnal inhabitant of mid-altitude forest in Usambara Mountains. Shelters under dead bark and in hollow

trees, feeding on sleeping chameleons and possibly frogs. Nothing known of its reproduction. A quiet, shy snake. May bite when handled, but not dangerous.

WHITE-LIPPED SNAKE
Crotaphopeltis hotamboeia 60–75cm

A medium-sized snake with *glossy black* temporal region and *white flecks* on an *olive green body. Lips white.* Has large eyes with vertical pupils in broad head. Lives in marshy areas and feeds at night. Widespread throughout most of region. Lays 6–19 eggs. Bad-tempered, and gives a dramatic threat by flattening the head to accentuate the white lips. Strikes readily, but is harmless. Can inflict deep punctures with long back fangs, but uses them mainly to 'pop' the toads and frogs that form its main diet.

EASTERN TIGER SNAKE
Telescopus semiannulatus 60–100cm

A distinctive, nocturnal, *thin-bodied* species with a *broad head* and *large eyes*. Has *19 scale rows* at midbody. Dull

orange body bears *22–50 dark blotches* that are larger on forebody. Restricted to the southern regions. Although mainly terrestrial, regularly climbs into trees or onto roofs to search for food. Small birds and lizards form main diet, although mice and even bats are also caught. Lays 6–20 eggs in moist leaf litter. When disturbed, bites readily and often, but the mild venom is not dangerous.

LARGE-EYED CAT SNAKE
Telescopus dhara 80–130cm

A distinctive nocturnal species with a *thin neck, wide head, large eyes* and long tail. Body *triangular* with *19–25 scale rows at midbody*. Body colour may vary from orange to black, and is

usually uniform but sometimes has narrow, pale crossbars. Restricted to the drier northern regions, extending to Arabia and Mauritania. Although mainly terrestrial, may climb into trees and shelter in bird nests. Eats small birds, lizards, mice and even bats. Lays 5–20 eggs. When disturbed, will hiss and strike readily, but is not dangerous.

SPOTTED BUSH SNAKE
Philothamnus semivariegatus 80–130cm

A beautiful diurnal snake that hunts among bushes on rocky ridges or along river courses. *Body slender* with a *long tail. A lateral keel* runs on each side of *belly and tail. Green body* has *black spots or crossbars* on the forepart and may become grey-bronze towards tail. Inhabits mainly savanna, but also forest edges. An expert and speedy climber, it pursues lizards and tree frogs. Lays a small clutch of 3–12 eggs. When confronted, inflates the throat to expose *vivid blue skin* between scales and strikes readily, but despite this bluff, is harmless.

WESTERN GREEN SNAKE
Philothamnus angolensis 80–120cm

A large and robust snake with smooth scales in *15 midbody rows.* Ventrals are *not keeled.* Body *bright emerald green* with white lower edge to each scale, visible when

body inflates. Belly greenish. Feeds on frogs, lizards and nesting birds. Lays up to 16 elongate eggs, and females sometimes nest together, when as many as 85 eggs have been found in rotting vegetation. Locally restricted to the south and west of region. Harmless, but inflates the throat in threat.

SPECKLED GREEN SNAKE
Philothamnus punctatus 70–100cm

Medium-sized, *narrow-headed*, with a prominent *raised eye ridge*. Has *sharp keels* on both *ventrals* and *subcaudals*. Usually has 15 scale rows at midbody. Body bright green or yellow-green, sometimes with fine black speckles and blue head. Active, searching among trees and bushes for lizards that form its main diet. Will also take frogs and small birds. Found mainly in south of region, with scattered records in Kenya. Lays up to six elongate eggs. Often confused with young green mamba, but is completely harmless.

GREEN WATER SNAKE
Philothamnus hoplogaster 60–93cm

A medium-sized, *round-headed* species that *lacks keels* on ventrals and subcaudals and has *black skin* between the scales. Usually has *15 scale rows at midbody*. Body bright green or grey-green, often with yellow throat. Some juveniles have black crossbands on forebody that may persist in some adults. An active swimmer that favours pans and backwaters, where it hunts small frogs. Found mainly in south of region, with scattered records in northwest. Lays up to eight elongate eggs. This gentle species rarely attempts to bite and does not inflate the throat in threat. Harmless.

BATTERSBY'S GREEN SNAKE
Philothamnus battersbyi 60–90cm

A medium-sized, moderately elongate species. Head has *large eyes* but *no raised eye ridges*. Usually has *15 scale rows at midbody*, and *subcaudals are not keeled*. Body colour ranges from bright green to dull grey-green, with small white base to each scale, only visible when body inflates. An active snake, searching among trees and bushes for the frogs that form its main diet. Will also eat lizards and even fish. Found mainly in the north of the region, occupying mid-altitudes in Kenya and adjacent Tanzania, extending elsewhere to Ethiopia. Lays up to 11 elongate eggs, sometimes in communal sites. Often confused with young green mamba, but is completely harmless.

USAMBARA GREEN SNAKE
Philothamnus macrops 50–80cm

A small species that has *large eyes* but *no raised eye ridges*. The subcaudals are keeled, but not the ventrals. Has only *13 scale rows at midbody*. Body colour highly varied, ranging from uniform green to red-brown, often with irregular crossbars and scattered blotches and spots. Diurnal, searching among coastal trees and bushes. Diet unknown, but probably includes frogs and lizards. Restricted to coastal Tanzania, possibly extending into northern Mozambique. Lays up to 14 elongate eggs.

BLACK-LINED GREEN SNAKE
Hapsidophrys lineatus 60–110cm

A large green snake with *strongly keeled body scales* in *15 rows at midbody*. Head has rounded snout and *large golden eyes*, while *tail is short, less than 33% of body length*. Tongue light blue with a black tip. Head and body emerald green, with each body scale *edged in black*, giving a finely lined appearance. Belly pale yellowish-green with *keeled ventral scales*. Inhabits lowland and riverine forest, hunting frogs on marshy ground. Lays small clutch of 3–4 large eggs. Just enters the region in the northwest. May bite, but is harmless.

EMERALD SNAKE
Hapsidophrys smaragdina 60–110cm

A large green snake with *strongly keeled body scales* in *15 rows at midbody*. Head long, with *large yellow eyes*. *Tail also long, more than 33% of body length*. Head and body *emerald green*. A prominent *black line* runs through eye on each side of the head. Lower lips and throat often bright yellow and belly pale yellowish-green with *keeled ventral scales*. Inhabits woodland, secondary forest and riverine forest west of Lake Victoria. An arboreal snake, it climbs well and hunts lizards and frogs in thick vegetation. Lays a small clutch of 3–4 large eggs. Considered harmless.

COMMON EGG EATER
Dasypeltis scabra 55–90cm

Body scales *strongly keeled*. Dirty grey, usually with *numerous (42–79) dark blotches* and a single *prominent 'V'-shaped mark on neck* that mimics coloration of some small adders. Bird eggs form sole diet. Undergoes long fasts out of nesting season. Lays up to 25 eggs. When disturbed, puts on a dramatic display, pretending to be dangerous, gaping mouth wide to reveal *black lining*, and striking readily. May also make a hissing noise by rubbing enlarged and keeled flank scales against one another. Egg eaters, however, are harmless, as they have *minute teeth*. Widespread in region, being absent only from aird parts of northern Kenya.

MONTANE EGG EATER
Dasypeltis atra 50–100cm

Similar to previous species, with *strongly keeled* body scales, *small head* and *rounded snout*. Body colour is varied and includes an all-black phase (usually restricted to forests), a uniform tan to reddish-brown phase (usually restricted to grasslands), and a patterned phase with *numerous (85–107) fine, blotched bars* (found in savanna). Bird eggs form sole diet. Considered harmless.

EAST AFRICAN EGG EATER
Dasypeltis medici 50–90cm

A secretive, but beautiful snake with a *small head*, largish eyes, *round pupils* and *strongly keeled body scales*. Body warm pink or pinkish-grey, with dark stripe along backbone that is interrupted by small white bars that extend as narrow dark bars onto light flanks. About *3–7 narrow 'V'-shaped marks occur on neck*. Northern specimens become patternless. Mouth lining *pinkish*. An arboreal species, restricted to coastal and riverine forest. Lays 6–28 eggs. Considered harmless.

WESTERN FOREST EGG EATER
Dasypeltis fasciata 55–90cm

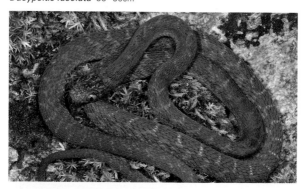

Body scales of this egg eater *strongly keeled*. Dirty grey to brown body has *numerous dark blotches* and a single *prominent 'V'-shaped mark on neck*. Bird eggs form sole diet, undergoes long fasts out of nesting season. A snake of the Congo and Upper Guinea forests, entering region only in western Uganda. Elsewhere is reported to lay 5–9 eggs.

HOOK-NOSED SNAKE
Scaphiophis albopunctatus 90–130cm

Steve Spawls

Hook-nosed snake looks superficially similar to a beaked snake (p.43) because they have evolved to live similar lifestyles, not because they are closely related. *A large, thick-bodied* snake with a conspicuously *hooked snout, small eye* and *round pupil. Parietal scales* on back of head are *fragmented.* Body *grey*, often with pink to rust-red infusion on head and flanks and grey region along backbone. Inhabits savanna, entering coastal thicket from Ghana to Angola, with disjunct populations in central arid region. May lay up to 30 eggs. Rodents form main diet and are crushed against walls of their burrows. When disturbed, strikes readily with a *gaping, black-lined mouth*, but does not bite and is harmless.

POWDERED TREE SNAKE
Toxicodryas pulverulenta 80–130cm

A slender, graceful tree snake with a narrow neck, broad head and large eyes. *Lips speckled.* The body has *vertebral ridge* and smooth body scales in *19 midbody rows.*

Juveniles and adults *pinkish-brown to reddish.* Has irregular bars, often rhombic in shape and pale-centred. *Fine black speckling* covers body and belly, which is pale pink, with dashed dark line on edges of ventrals. An elegant nocturnal climber in moist forest of Congo Basin, reaching eastern limit in Uganda. Feeds on small rodents and lizards. Lays small clutch (2–5) of large eggs. Inflates body in threat, but not known to be dangerous.

BLANDING'S TREE SNAKE
Toxicodryas blandingii 150–240cm

Olivier Pauwels

Marius Burger

Top: *Adult male* Bottom: *Subadult*

A very large tree snake with a narrow neck, broad, powerful head and wide gape. *Lips yellow with black lines* between scales. Smooth body scales in *21–23 midbody rows*. Backbone prominent, giving *triangular cross section*, and tail long and thin. Juveniles brown with irregular bars that are sometimes diamond-shaped and pale-centred. Bars fade with age, but usually visible in large females. Males become *glossy black above and below*. Inhabits rainforest patches in western regions. A powerful climber, active at night and eats wide range of small vertebrates. Lays up to 14 eggs. In threat inflates the body, coils forebody and gapes widely. Will bite readily, and venom has caused nausea and headaches, although fatalities unknown. Treat large adults with care.

JACKSON'S TREE SNAKE
Thrasops jacksoni 120–230cm

A large arboreal snake. Could easily be confused with a boomslang (p.74), due to its habits and oblique, *strongly keeled body scales* along back. However, is uniform black in colour, has *smaller eyes*, and *smooth lateral body scales*. Juveniles have greenish heads and brown-and-black mottled bodies. Locally restricted to Great Lakes region, extending elsewhere to northern Zambia, southern Sudan, and Central African Republic. Lays 7–12 eggs. Eats mainly chameleons and fledgling birds. Like boomslang, inflates the neck in threat. Venom has not been studied, and no case histories reported. However, snake should be treated with care.

SPLENDID TREE SNAKE
Rhamnophis aethiopissa 100–150cm

A large tree snake with *enlarged scales* along backbone. Head distinct, with *very large eyes*. *Body scales* are *always smooth*, elongate and in *15–17 midbody rows*. Head olive, sometimes with black-edged

scales on crown. Body green, with black-tipped or bordered scales. Tail black with green stripe on each scale row, while chin and throat yellow-green, otherwise pale green below with dark median line below tail. Inhabits forest of the Congo Basin, reaching eastern limit in Uganda. A graceful climber in lowland and riverine forest, hunting lizards and frogs. Lays up to 17 large eggs. When threatened, inflates the neck, but is not venomous.

DAGGER-TOOTH SNAKE
Xyelodontophis uluguruensis 80–130cm

Only described in 2002, this *thin, elongate* snake has a *lance-like head*, large eyes with *pear-shaped pupils*, *single nasal scale* and *divided loreal*. Has *feebly keeled body scales* and *long tail*. Top of head bronze, forebody finely barred in yellow and black and skin between neck scales blue. Belly dark with light speckles. Restricted to montane evergreen forest on Uluguru Mountain, Tanzania. In threat, holds body stiff and inflates throat to appear bigger. Bites readily, but venom toxicity unknown. Treat with caution, as close relatives have fatal bites.

EASTERN TWIG SNAKE
Thelotornis mossambicanus 90–140cm

The *very thin, elongate body, lance-like head* and cryptic coloration make this snake unmistakable. Has large eyes with *keyhole-shaped pupils*. Body scales *feebly keeled*, tail *very long*. Body resembles a twig: grey-brown with black and pink flecks and series of diagonal pale blotches. Crown of head *green to pale brown*, often *speckled with black*. Restricted to southern savannas, extending south to Mozambique. Completely arboreal and hunts lizards and small birds, swallowing them while hanging downwards among branches. Lays 4–18 elongate eggs in rainy season. Has potent venom that causes death as a result of internal bleeding.

USAMBARA TWIG SNAKE
Thelotornis usambaricus 80–130cm

Steve Spawls

Very similar to previous species except that crown and sides of head are *bright green (sometimes yellow), never speckled with black.* Body mottled grey, green and brown, with *black crossbars on the forebody.* Has 156–169 *ventral scales.* Completely arboreal and is restricted to montane and coastal forest in scattered localities. Biology similar to that of Eastern Twig Snake. Occurs on Eastern Usambara and adjacent mountains, with isolated coastal populations, including on Vamizi Island, northern Mozambique. Venom presumably potent, but no case histories are known.

FOREST TWIG SNAKE
Thelotornis kirtlandii 90–140cm

Very similar to previous species except that crown and sides of head are *always bright green (never yellow),* Body mottled grey, green and brown, with *black crossbars on forebody.* There are

162–187 *ventral scales.* Completely arboreal and restricted to closed-canopy forest in scattered localities. Biology similar to that of Eastern Twig Snake. Enters the western region, with isolated populations in Eastern Arc Mountains of Tanzania. Venom potent and, although no fatalities known, is probably deadly. Should be treated with care.

BOOMSLANG
Dispholidus typus 110–180cm

Top: *Adult female* Above: *Juvenile*

One of Africa's most characteristic snakes. Widespread, but absent from arid scrubland and closed-canopy rainforest. Easily distinguished by its short head, *very large eyes* and *oblique, strongly keeled body scales*. Coloration highly variable. Juveniles have bright emerald eyes, white throats and twig-coloured bodies with blue throat skin. Females remain drab olive, while males may become bright green all over, or even green and black with yellow bellies. Lays up to 25 eggs. A dangerous, but shy, diurnal snake that hunts chameleons and small birds. When disturbed, inflates the throat and will bite readily. Venom prevents blood clotting, and death may follow in 1–3 days. Shy and retreats when encountered, so not dangerous if left alone.

LIZARDS   Suborder Sauria

SKINKS   Family Scincidae

A large, diverse Old World family, with limited presence in New World. Skinks have *shiny, overlapping scales* (like fish scales) with an internal bony layer (osteoderm), flexible armour that allows them to burrow underground or live in rock cracks. Numerous species have reduced limbs or have lost them completely. Head has *large, symmetrical scales* and tail can be shed and regenerated.

CURROR'S LEGLESS SKINK
Feylinia curruri 18–33cm

Marius Burger

A *robust legless* skink with a *wide head, pointed snout,* and *no external ears or eyes.* Body has *20–28 smooth midbody scales,* and *elongate tail has blunt tip.* Ocular scale is in *contact with third upper labial.* Body shiny black, sometimes silvery grey (especially just before skin is shed), above and below. Rarely seen above ground as it lives in leaf litter or loose soil, searching for the temites and insect larvae that form its diet.

FOUR-TOED BURROWING SKINK
Sepsina tetradactyla 10–15cm

A small, *elongate* skink with a *rounded snout, lidded eyes,* elongate tail with blunt tip and *minute feet,* with only *4 toes on front feet.* Body bronze-brown, pinker on belly, and tail has *blue infusion.* Rare, but recorded in scattered localities in southern and western Tanzania. Burrows in leaf litter and in loose soil around tree roots, particularly on large vegetated termite nests. Females may contain up to four eggs, but it is not known whether they are laid or retained to develop within the mother.

COASTAL RAG SKINK
Cryptoblepharus africanus 8–10cm

Small, slender skink with large eyes, each covered by *immovable transparent spectacle*. Body covered with smooth, close-fitting scales in *26–29 midbody rows*, and tail tapers to fine point. Blackish-bronze back has numerous pale spots on flanks and legs. Restricted to coastal rock outcrops throughout much of Indian Ocean region. A diurnal species that forages on intertidal rocks to catch small crustaceans and even small fish. Swims easily, and dives into shallow pools to escape predators. Lays 1–2 eggs in sand.

WAHLBERG'S SNAKE-EYED SKINK
Afroablepharus wahlbergi 8–10cm

A small skink, easily identified by *immovable eyelids* and grey-bronze body, which may *have six fine dark lines*. Has *smooth body scales* in *24–26 rows at midbody*. Belly greyish-blue, except in breeding males, in which it turns pinkish-orange. Numerous isolated populations known from Ethiopia to South Africa, suggesting that this is a complex of cryptic species. Scuttles among grass roots and rotting logs, feeding on termites and small insects. They are short-lived, surviving for only 10–14 months. Females lay 2–6 eggs in rainy season.

SUNDEVALL'S WRITHING SKINK
Mochlus sundevalli 15–18cm

A small, fat-bodied skink that wriggles among leaf litter and loose sand. Has lidded eyes, smooth scales in *24–30 midbody rows*, and *fat tail*. Feet are small, but all have *five toes*. Body uniform bronze, or appears speckled, due to small spot on each scale. Belly is paler. Common throughout savanna regions, where it feeds on termites and insects around rotting logs. Lays 2–6 soft-shelled eggs under stone or in old termite nest. Its tail was once prized as a cure for snakebite.

PETER'S WRITHING SKINK
Mochlus afrum 16–23cm

Above: *Yellow-bellied form*

Difficult to tell apart from previous species except by size, this large, fat-bodied skink has *lidded eyes*, smooth body scales in *26–28 midbody rows*, and long, fat tail ending in spine. Feet are small, but all have *five toes*. Bronze body speckled with dark and pale spots, while belly is paler and sometimes yellow. Common throughout coastal and eastern savanna regions, where it burrows in sandy soil, feeding on termites and insects found around rotting logs. Lays up to seven soft-shelled eggs in loose, sun-warmed sand.

RED-SIDED SKINK
Lepidothrys hinkeli 20–30cm

Reto Kuster

A beautiful skink restricted to forest fragments in Great Lakes region. Unfortunately very shy and rarely seen. Has big, blunt head and *stout, shiny* body with *34–38 midbody scale rows*. *Limbs short*, but strong, each with five toes. Body speckled bronze-brown above, with *red bars on flanks* speckled with white and black. Has series of irregular black blotches on sides of neck and in forelimb region. Sides of head and lips are red. Juveniles have blue and black tails that fade to black with age. Usually found in holes among buttress roots of big trees in swampy areas and possibly nocturnal, but probably favours dusk and early morning. Inactive for long periods in its burrow, but emerges to feed on insects. Lays eggs, but little known of its reproduction. Relatively slow and docile, but will give strong bite.

RWANDA FIVE-TOED SKINK
Leptosiaphos graueri 12–20cm

Chris Kelly

An *elongate skink* with *very small five-toed limbs* (rarely four-toed on front feet). *Long tail is almost twice body length.* Head has short, rounded snout and lidded eyes. Body and tail metallic red-brown above. Flanks

heavily speckled with black and white that may form dark bars on midbody. Belly grey-white and salmon-pink towards tail. Found in wet forest in western regions and wriggles among roots and ferns in search of small insects. Lays only two small, elongate eggs.

SHORT-NECKED SKINK
Trachylepis brevicollis 18–32cm

Large, *heavy-bodied,* with short head. Body has *2–3-keeled scales in 30–34 midbody rows.* Juveniles black with yellow bars on flanks. With age, number of yellow spots increases and body colour lightens. Adult females

Colin Tilbury

brown with irregular brown blotches, while males have brown and black stripes, and white spots on forebody. Restricted to northern savannas from Ethiopia to northern Tanzania. Mainly terrestrial, but does climb onto rocks and trees, where it lives in diffuse groups. Gives birth to live young.

BICOLOURED SKINK
Trachylepis dichroma 18–32cm

Large, *heavy-bodied,* with short head. Body scales have *2–3 keels* and are in *34–38 midbody rows.* Juvenile's back

Top: *Adult* Above: *Juvenile*

has network of black with golden centres, fine black and white bars on flanks, and finely black-banded tail. Adult male and female coloration similar to that of juvenile, but with less conspicuous network. Lips golden-yellow. Breeding males develop dark backs with orange-red bellies. Widely distributed in open *Acacia* savanna of central Kenya and adjacent Tanzania. In Serengeti region lives in small groups and basks on old termite nests. Gives birth to 2–3 large young.

SPECKLE-LIPPED SKINK
Trachylepis maculilabris 18–28cm

A large arboreal skink with *pointed snout* and *long tail*. Body scales have *5–8 keels* and are in *29–35 midbody rows*. Body colour varies regionally, but distinguished by *yellow eyelids* and *black-and-white barred lips*. Body brown, often darkly speckled, and sides of neck speckled with white. May have orange bar on flank (Kenyan coast). Climbs in low bush, sheltering under bark or in hollow trunks. Lays 6–8 eggs and feeds on small insects.

ALPINE MEADOW SKINK
Trachylepis irregularis 15–24cm

A heavy-bodied, short-tailed, terrestrial skink. Body scales have *2–5 keels* and are in *31–34 midbody rows*. Back dark, with fine yellow lines (which may be broken into spots) and finely speckled flanks. Pinkish belly may have irregular black stripes. Soles of feet also pinkish. Restricted to high-altitude grassland over 3,000m on the Aberdares, Mt Kenya and Mt Elgon. Sheltering in grass tussocks and under stones, is active only on sunny days. Gives birth to a few young. Eats insects, including large ants.

FIVE-LINED SKINK
Trachylepis quinquetaeniata 15–25cm

Colin Tilbury

A large, *rock-living skink* with a stout body and limbs, and longish tail (adult male shown here has lost most of his tail, but it is regenerating; see black tip). Body scales have *three keels* and are in *32–42 midbody rows*. Juveniles have dark bodies with *five yellow-white stripes* and electric blue tails. These persist in large females, but tails of breeding males becomes reddish-brown.

Dorsal body stripes also disappear and body develops white flecks. Side of neck is black with white spots and bars and a yellow stripe. Occurs in scattered populations in northern regions. Forms large colonies with dominant males and lays 3–10 eggs

RAINBOW SKINK
Trachylepis margaritifera 18–30cm

Left: *Breeding male*
Right: *Juvenile*

This skink is related to previous species, but *tri-keeled body scales* are in *38–52 midbody rows*. Females and juveniles have dark bodies with *three bluish-white stripes* and electric blue tails. Breeding males become golden-brown, with scattered pearly-white spots and greenish tail. Can be found in isolated populations in southern Kenya

and central and eastern Tanzania, thence to South Africa. Rock-dwelling. Dominant males fight over territories, but allow the blue-tailed juveniles and females to remain unchallenged. Lays 6–10 eggs in rainy season that take about two months to hatch.

VARIABLE SKINK
Trachylepis varia 12–16cm

A small skink with characteristic *bright white lateral stripe* (yellow in Aberdare specimens). Has *tri-keeled* scales in

27–36 midbody rows, and *scales beneath toes are keeled*. Dark reddish-brown back may have black spots and additional pale stripes. Restricted mainly to the eastern regions, thence to South Africa. Hunts in broken ground, climbing onto boulders and fallen trees. Insects, which form main diet, are captured after a short dash from cover. Gives birth to 3–10 young.

STRIPED SKINK
Trachylepis striata 18–25cm

Widespread throughout most of region, these rather dull skinks tame readily and are common around houses. Scales have *3–7 keels* and are in *32–43 midbody rows*. Dark brown to black body has *bold, golden-white dorsolateral stripes,* and flanks are usually speckled. Belly cream-white with darker speckling, particularly on throat. Females give birth to about 3–9 young throughout year.

BOULENGER'S SKINK
Trachylepis boulengeri 15–28cm

A medium-sized, slender skink with a very long tail (almost *twice body length*). Body scales in *28–32 midbody rows* and usually have *7–9 keels*, but juveniles may have only three keels, increasing to 11 with age. Body unpatterned, *dull warm bronze* to grey-brown, sometimes with faint dark flecks. Chin, throat and belly yellowish. Restricted to southern Tanzania and occurs southwards to Zimbabwe. Diurnal, foraging in thick vegetation around marshes and along drainage lines. Feeds on insects. Lays 3–5 eggs. Often found sleeping coiled in long grass, sometimes overhanging water.

TREE SKINK
Trachylepis planifrons 22–35cm

A *large, stout* skink with a tail nearly *twice as long* as body. Body scales have 3–5 weak keels in 25–32 midbody rows. Body grey-tan with dark lateral stripe running from snout to hindlimbs, below which flank is white with fine stippling. Colour bolder in northern populations. Found in scattered populations in wooded savanna, mainly in eastern regions (but rarely to coast). Elsewhere from Ethiopia to Zambia. Arboreal, lives in tree hollows and hunts large insects. Lays eggs, but there are few other details of reproduction.

OLD WORLD LIZARDS Family Lacertidae

Usually small and slender with long tails. Have small granular scales on back, larger, squarish scales on belly, large symmetrical scales on head and can shed and regrow tails.

BOULENGER'S SCRUB LIZARD
Nucras boulengeri 12–18cm

 Cryptically coloured, terrestrial, with a blunt snout, *smooth scales* beneath toes, and tail almost *twice as long as body*. Throat

has *collar* of enlarged scales.
Grey-olive with darker spots, faint dorsal stripes, and white spots on flanks. Belly white. Juveniles have reddish tails and pale back stripes. Found in the central parts. Shelters in burrow, emerging to feed on insects, particularly winged termites. Lays eggs, but clutch details are unknown.

SPEKE'S SAND LIZARD
Heliobolus spekii 10–18cm

 Small, slender, terrestrial. Body covered in *fine granules*. Has *curved collar* of enlarged scales beneath neck.

Scales beneath digits *keeled*. Belly scales rectangular, in *six rows*. Back brown with even black spots between fine golden stripes. Limbs darker, with pale spots. Belly white. Widespread in savanna and semi-arid regions. Sits in shaded spots and ambushes small insects that land nearby. Lays 4–6 eggs in sand beneath scrub.

GREEN KEEL-BELLIED LIZARD
Gastropholis prasina 25–40cm

A large, sleek, *verdant green* lizard with a *very long prehensile tail*, more than twice body length. Limbs well developed and toes all have strong claws to aid climbing. Narrow head with *well-developed collar* on neck. Dorsal scales small and granular, but those on *light green belly are wide, in 8–12 rows*, and *keeled.* Found in isolated forest pockets in Eastern Arc Mountains and a few coastal localities in Kenya–Tanzania border region. Arboreal and diurnal, hunting insects and even small lizards among branches. Shelters in tree hollows, where it lays up to five eggs. Threatened due to restricted range and habitat loss.

STRIPED KEEL-BELLIED LIZARD
Gastropholis vittata 22–35cm

Very similar to previous species in habits, but smaller and with different coloration. Body *dark brown with two white dorsal and lateral stripes.* Belly *cream. Collar white.* Found in

coastal forest patches from Diani Beach to Dar es Salaam with scattered records to the south, including northern Mozambique. Biology is probably similar to that of the Green keel-bellied Lizard.

SOUTHERN LONG-TAILED LIZARD
Latastia longicauda 20–40cm

A large, *very long-tailed* terrestrial lizard that hunts in semi-arid northern regions, extending through the Sahel to Senegal and north to Yemen. Has *serrated collar* beneath neck, *keeled* scales beneath digits, and *rectangular belly scales in 6–8 rows*. Body varies from red-brown to grey, with fine pale dorsal stripes and numerous irregular black bars. Flanks may have blue, black-edged spots (ocelli), while tail is unmarked. A very common and conspicuous lizard, wary if approached, and rushes off at great speed, disappearing down burrows or into thicker bush. Lays eggs but clutch details unknown.

JOHNSTON'S LONG-TAILED LIZARD
Latastia johnstoni 12–20cm

A fast, active lizard found in sandy savanna flats, dry riverbeds and around rock outcrops. Body covered in *fine granules*, with large *white rectangular belly scales*. Toes have *keeled scales* below and *tail much longer* than body. Juveniles have prominent stripes and an orange tail. Tail fades in adults, lateral blotches become more prominent and paired anterior vertebral stripes fuse to form less distinct pale vertebral stripe on hindbody. Restricted to central Tanzania, thence to Malawi. Dashes from cover to seize insects. Lays small clutch in soft sand.

EASTERN BLUE-TAILED GLIDING LIZARD
Holaspis laevis 8–13cm

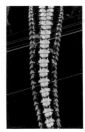

A small, agile, *brightly coloured tree* lizard. Body *very flattened* with series of *enlarged scales* down spine. Feet have long digits and tail fringed with row of *flattened yellow lateral spines* that increase its width, allowing lizard to take long, gliding leaps between branches, or to glide to ground without injury. Body black with broad pair of golden dorsolateral stripes that fuse on tail base and become blue. *Single finer golden stripe* occurs on flank, while central golden stripe runs from snout to back of head. Found in scattered coastal forest from Kaya Jibana forest, southern Kenya, to central Mozambique. Lays pair of eggs under tree bark.

ROUGH-SCALED DESERT LIZARD
Meroles squamulosa 17–20cm

Recognised by *small head* and *small, strongly keeled body scales*, this active lizard hunts in sandy clearings for grasshoppers and termites. Cryptically coloured. Buff-brown body has narrow dark crossbands or blotches with long rows of pale spots. Restricted to extreme southern Tanzania, thence to South Africa. Very short-lived, growing to maturity in 8–9 months and dying soon after breeding, at only 12–13 months. Digs branching burrows in soft sand at base of bushes. Lays 8–12 eggs Apr–Jun.

ANGOLAN ROUGH-SCALED LIZARD
Ichnotropis bivittata 17–20cm

A small terrestrial lizard that *lacks collar*, has small, *strongly keeled body scales*, and *keeled scales beneath toes*. Back brown with pale-edged black dorsolateral stripe and red flanks. Lips and lateral neck stripes chrome yellow, and belly white. Juveniles duller in colour, but have round white spot above shoulder. Enters southwestern savanna regions, extending to Zambia and Angola, and is active diurnal hunter of small insects. May also be short-lived (see previous species), but its biology is poorly known.

WESTERN ALPINE MEADOW LIZARD
Adolfus marasaensis 12–18cm

A small terrestrial lizard with a *smooth collar* beneath neck, *keeled body scales, smooth scales* beneath toes, and *longish tail*. Back red-brown to lime green, with thin dark line along spine and pale edge to darker flanks. Belly ranges from vivid orange to bluish. Restricted to high-altitude moorland on the Aberdares and Mt Elgon. Forages among grass clumps and on fallen logs, but secretive and easily overlooked. Is presumed to lay eggs, but its biology is poorly known.

JACKSON'S FOREST LIZARD
Adolfus jacksoni 15–25cm

Similar to previous species, but with smaller *keeled body scales in 37–48 midbody rows*. Body heavily speckled, with pale dorsal band and reddish flanks. Scattered yellow spots border yellowish belly. Inhabits clearings in highland forest, clambering onto rock faces and logs, hunting small insects. Lays up to five eggs, often in rock crevices, where communal egg-laying sites may contain hundreds of old and new eggs. Eggs take up to two months to develop.

PLATED LIZARDS Family Gerrhosauridae

Endemic to Africa and Madagascar. Diurnal oviparous lizards, most having stout bodies, long tails and well-developed limbs. Body has prominent lateral fold and scales are rectangular with bony plates.

ELLENBERGER'S LONG-TAILED SEPS
Tetradactylus ellenbergeri 20–30cm

M Menagon

An elongate *snake-like* lizard. *Tail ¾ of total length. Forelimbs absent* and *hindlimbs* consist of *minute vestigial spikes. Head blunt* with *smooth scales.* Body has prominent *lateral body fold* and *polished keeled* scales. Body and tail olive brown, paler below. Restricted to moist savanna in southern Tanzania, and west to Angola. 'Swims' rapidly through short vegetation, chasing grasshoppers and other insects. Shy, easily shedding tail in defence. Lays small clutch of elongate eggs.

ROUGH-SCALED PLATED LIZARD
Broadleysaurus major 30–40cm

Solid, with a *short head, large eyes* and *long, strong tail.* Dorsal scales large and very rough. *Belly scales* in *10 rows.* Rounded brown body often lighter at the front. Gold tips to scales on hindbody, and flanks give speckled appearance. Chin and throat usually pale cream, sometimes light blue or pink. Restricted to the southeast, thence to South Africa. Favours soil-filled cracks in wooded rock outcrops, as well as vegetated termite hills. Feeds on large insects, but also soft fruit and flowers, and even other small lizards. Lays 2–6 large, soft-shelled eggs beneath a rock or dead log. Common, but rarely seen.

YELLOW-THROATED PLATED LIZARD
Gerrhosaurus flavigularis 25–35cm

Small head, long tail, prominent *lateral body fold,* and *dark-edged bright yellow lateral stripe* on each flank make this elegant medium-sized lizard unmistakable. Scales on soles of feet are *smooth.* Breeding males usually have rust-red snouts and yellow throats, although blue-throated specimens also known. Widespread in eastern regions, extending into central Kenya, elsewhere to South Africa. Terrestrial, living in small burrows at base of bushes, emerging occasionally to feed on insects. Lays 4–8 eggs at start of wet season.

EASTERN BLACK-LINED PLATED LIZARD
Gerrhosaurus intermedius 30–45cm

A large plated lizard with a prominent *lateral body fold* and long tail that is *twice body length.* Belly scales in *eight rows,* and *scales on soles of feet keeled.* Brown body has speckled appearance, with

fine yellow lateral stripe and *red flanks.* Restricted mainly to eastern Tanzania, thence to South Africa. Lives in burrow dug at base of shrub in moist savanna or coastal thicket. Uses powerful jaws to tackle large grasshoppers, beetles, millipedes and even snails. Lays up to nine eggs in burrow beneath log or rock and these take 70–80 days to develop.

GIRDLED LIZARDS Family Cordylidae

Endemic to Africa. *Rectangular body scales* are keeled and arranged in regular rows (girdles). Body usually flattened and almost box-like in cross section, with lateral body fold.

GRASS LIZARD
Chamaesaura miopropus 35–55cm

Another diurnal, snake-like, elongate lizard. *Elongate head has rough scales. Minute forelimbs,* and short hindlimbs with *single clawed digits.* Body scales

rough and elongated into spines. Body and tail light brown with two darker lateral stripes. Restricted to moist savanna in southeastern Tanzania, and west to Zambia. 'Swims' like a snake, moving rapidly through vegetation in pursuit of grasshoppers and other insects. Bears 6–8 young during wet season.

UKINGA GIRDLED LIZARD
Cordylus ukingensis 8–13cm

A small, *very spiny* girdled lizard, with a *cylindrical body* and *very short tail.* Body and belly scales spiny, with whorls of very spiny scales on tail. Body light golden-brown, mottled

with dark brown speckles. Belly paler, with grey flecks. Terrestrial, living in small burrows in grassland. Emerges during the day and hunts small insects. Gives birth to a few small young. Very poorly known. Restricted to the Ukinga and Udzungwa mountains.

MAASAI GIRDLED LIZARD
Cordylus beraduccii 10–17cm

Large girdled lizard, with a *flattened body* and *prominent lateral body fold*. Tail about as *long as body*, and ringed with whorls of spiny scales. Head dull brown above, lips and throat white, and red-brown body has darker blotches and yellow flecks. Belly buff-coloured. Restricted to northern Tanzania and adjacent Kenya. Shelters in low rock outcrops in arid grassland, often sharing rock crevices with Pancake Tortoises (p.123). Like all girdled lizards, gives birth to a few small young.

TROPICAL GIRDLED LIZARD
Cordylus tropidosternum 13 16cm

This arboreal girdled lizard is *round-bodied* but retains weak *lateral body fold*. Has *very rough scales* and lichen-coloured back, with lighter flanks and belly. Dark lateral stripe extends from neck to

groin, and upper lip speckled cream. Shelters in hollow logs and under bark, and feeds on moths, spiders and winged termites. Rarely ventures far from safety of its retreat and lays down large fat reserves to survive through long dry season. Bears 2–4 large young. Restricted to Tanzanian coastal region, thence to Zimbabwe. Collected in large numbers for international pet trade, and also threatened by bush clearance for wood fuel.

MONITORS Family Varanidae

Monitors have long, flexible necks, well-developed limbs, strong claws, and long tails that cannot be shed or regenerated. *Head and body covered in small bead-like scales.* Tongue long, forked and snake-like. Occur throughout Africa and Australasia.

WATER MONITOR
Varanus niloticus 120–220cm

Very large, with *elongate head*. Tail *flattened with dorsal ridge* and is *much longer than body*. The *black-and-yellow-barred* juveniles are much brighter than adults. Forages along rivers for crabs and other aquatic organisms. Swims easily and is major predator of crocodile and terrapin nests. Juveniles eat mainly insects and small frogs. Lays up to 60 eggs, often in live termite nests, taking 4–6 months to develop. May give painful bite, or whip long tail in defence. A slightly larger species (*Varanus ornatus*) lives in Congo Forest and just enters Uganda.

ROCK MONITOR
Varanus albigularis 90–150cm

Similar to Water Monitor, but smaller, with a *shorter head and tail. Nasal region swollen. Drab, mottled* and usually sullied with old skin and ticks. Juveniles brighter, with *dusky blue-black throats*. Widespread in southeast, thence to South Africa. Wanders in rocky and savanna habitats, sheltering in tunnels and rock overhangs. Eats large insects and millipedes, but also small tortoises and lizards. Escapes into trees or rock cracks, or will bite and lash with tail if cornered. Lays up to 50 eggs in moist bank, hollow tree, or rock crack. Bite painful, but not poisonous. Preyed on by eagles, honey badgers and large cobras.

AGAMAS Family Agamidae

A large, diverse family. East African species all very similar: heads have small, irregular scales, they are active, diurnal, and many species feed mainly on ants and termites. Males develop vivid breeding colours and defend territories. All lay soft-shelled eggs.

KENYAN ROCK AGAMA
Agama lionotus 20–35cm

Large, with a large *triangular head*. Body has *small, spiny, strongly keeled* scales. Limbs short and strong. Tapering tail *banded in blue and white* and cannot be shed or regenerated. In breeding season dominant males develop bright orange head and throat that is *uniform rufous with darker and brighter stripes*. Forebody purple-blue with yellowish dorsal line, hindbody and limbs blue-green. Females and juveniles cryptic. Widespread in Kenya, from coastal lowlands to central and northwest regions. Extends into Tanzania. Found in savanna on rock outcrops and trees. Social, in dispersed colonies with dominant males. Diurnal. Active by day, catching insects. Buries up to 10 eggs in holes or loose soil.

MWANZA ROCK AGAMA
Agama mwanzae 20–32cm

Large, active, with *conspicuous ear openings*, *clusters of spiny scales* on neck, and long, tapering tail. Breeding male beautiful, with vivid pink head and back, violet or blue-white line down spine, and bright blue limbs and tail. Females and juveniles drabber, with brown bodies and irregular dark crossbars. Social, living in colonies on rock outcrops in open savanna. Dominant males command best feeding and basking spots, but allow other lizards to share their retreats in rock cracks. Egg-laying, but clutch details unknown.

DODOMA ROCK AGAMA
Agama dodomae 20–35cm

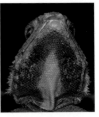

A medium-sized agama with *tail much longer than body*, a *weak crest* on nape, and *no spines around large earhole*. Body scales *keeled, of similar size* and in 80–90 rows. Males brightly coloured with *red head, dark blue to black throat* with *red centre, light purple to blue* body, sometimes with *pale vertebral stripe, light green-blue legs* and *narrowly banded blue-and-white tail*. Females smaller, with dull brown bodies and greenish heads speckled with white. Diurnal. Occurs in small dispersed groups on rock outcrops in semi-arid scrubland in central Tanzania. Eggs buried in loose soil at start of rainy season.

SPINY AGAMA
Agama armata 15–20cm

Small, with *short tail* and *small earholes*. Body scales *keeled* with *88–105 midbody rows*, and with *longitudinal rows of enlarged spiny scales* on back. Body grey to brown with 4–5 short crossbars. Throat has *irregular network* of dark lines, with *blue-black blotch at base*. Breeding males develop *bright blue throat*. Prefers open woodland from the Maasai Mara south to Lake Rukwa. Elsewhere to Mozambique and South Africa. Terrestrial and mainly solitary, although pairs may share short tunnel dug into sandy soil at base of a bush. Runs rapidly when disturbed and often uses rodents' tunnels as temporary shelters. Lays clutches of 9–16 eggs.

MOZAMBIQUE AGAMA
Agama mossambica 20-31cm

A large and slender agama with *uniform body scalation, 69–94* scales around the midbody, and a longish tail. A *thin black line runs from the eye to the ear.* Females and juveniles are mottled grey, white and tan, with a series of pale geometric shapes down the back. The breeding male develops a bluish head, and a prominent pale line along the spine. Restricted to southeastern Tanzania and from thence to Mozambique. Forages in leaf

litter, but climbs onto tree trunks in open woodland. Insects, particularly ants, form the main diet. Tolerant of people and may enter villages and homes. Clutches of up to 14 eggs are laid in humic soil, in a hollow tree or beneath a dead log.

MONTANE ROCK AGAMA
Agama montana 15–27cm

A small agama with a *long, thin tail* and *uniform body scalation.* Male's body olive brown, sometimes with scattered black scales and blue vertebral stripe. When breeding, males also develop ultramarine throat. Females golden-brown above, dirty white below. Endemic to Usambara and Uluguru mountains in eastern Tanzania. Mainly terrestrial, although sometimes found on rocks and trees. Avoids closed-canopy woodland, but

adapts to farm bush where it may live in fields and gardens. Mainly solitary. Females lay 6–10 eggs.

TREE AGAMA
Acanthocercus atricollis complex 20–35cm

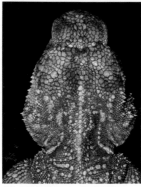

Adult breeding male (Mt Meru, Tanzania)

Large tree agamas have large heads and *conspicuous earholes*, but *lack dimpled occipital scale* on crown of head found in smaller agamas. Body covered in *keeled scales* with scattered *enlarged spiny scales* that may form *irregular bands* and may become bright orange when breeding males display. *Large black spot* occurs on each side of neck. Bold stripe, green-yellow in some populations, may develop along backbone of breeding males. In East Africa all tree agamas form part of *Acanthocercus atricollis* complex, but occur in diverse colours and body forms. At least 2–3 different species thought to be present. Large males very conspicuous when clinging to tree trunks or sitting on rock piles, nodding aquamarine heads in territorial displays. Have powerful jaws and will fight other males, biting base of tail, where scales are big and spiny, which helps prevent damage. Females and juveniles have cryptic, but attractive, lichen-coloured bodies. Tree agamas feed mainly on flying insects, but will eat beetles and even small vertebrates. When full of eggs and unreceptive to breeding, female develops orange-red blotches along spine. During wet season female comes down to ground to dig small hole in which she lays up to 15 eggs. This agama gapes widely in threat, and will give painful, but non-venomous bite.

CHAMELEONS Family Chamaeleonidae

Unmistakable lizards, adapted for life in trees. Restricted mainly
to Africa and Madagascar. Have compressed bodies and head
covered in *small granular scales*. Large, *turreted eyes can move
independently*. Feet have *toes bound in uneven bundles* and can clasp
thin branches. *Tail prehensile,* cannot be shed or regenerated. Food
captured using familiar *telescopic tongue*. Not venomous.

FLAP-NECKED CHAMELEON
Chamaeleo dilepis 20–24cm

The only chameleon
found in moist savannas
of southern regions.
Has *large, mobile ear
flaps* behind head, and
*crest of small white
scales on throat and
belly*. Flanks *lack
enlarged tubercles*.

Body colour may vary from green to pale yellow
or brown, often with white blotches and lines
on lower flanks. Gapes widely in defence, revealing
orange mouth lining, and flattens body while
rocking from side to side. Female digs long tunnel
and lays up to 57 small eggs, which may take
nine months to hatch.

SLENDER CHAMELEON
Chamaeleo gracilis 15–30cm

Colin Tilbury

A large chameleon, similar in shape and coloration to
previous species, but *lacks ear flaps* and is found in
northern woodlands. Often lives in acacia trees. When
disturbed, threatens by swelling throat to reveal orange
skin, opens mouth and hisses noisily. Female becomes
bloated with up to 44 eggs and may take several hours
to dig hole in ground in which to lay them.

GIANT EAST USAMBARA BLADE-HORNED CHAMELEON
Kinyonia matschiei 20–41cm

Left: Female Right: Male

A very large chameleon, with *two blade-like scaly horns* on snout, which are smaller in females. Crests poorly developed on back and throat, and absent from belly. Colour varies, but usually consists of irregular bars of green and yellow-grey often with *yellow belly patch* in males. Endemic to East Usambara Mountains. Lives in Afromontane forest, but also enters gardens. Clutches of 16–24 eggs laid in sun-warmed soil and may take nearly a year to develop. Hatchlings lack horns on snout.

KILIMANJARO BLADE-HORNED CHAMELEON
Kinyongia tavetana 18–23cm

A largeish chameleon, with *two blade-like scaly horns on snout*, which are absent in females. *Crests absent from body and belly*, and *tail is longer than body.* Colour varies, but usually consists of irregular bars of green and blue-grey, with broken pale dorsolateral line. Found in Mt Kilimanjaro, Mt Meru, Pare Mountain region, just entering Kenya, where it lives in montane forest, but adapts readily to gardens and coffee plantations. Clutches of 8–15 eggs are laid and take 4–5 months to hatch. Hatchlings lack horns on snout.

USAMBARA FLAP-NOSED CHAMELEON
Kinyongia tenuis 12–15cm

A small, slender chameleon, with simple, *laterally flattened, scaly horn* on snout in *both sexes. Crests absent* on body and belly, and *tail slightly longer* than body. Main body colour olive green with *patches of light blue*, particularly in two bands on flanks and on lips and horn of breeding males. Restricted to Shimba Hills (Kenya) and eastern Usambara Mountains. Lives in trees on edge of lowland forests. A clutch of five eggs has been recorded.

UTHMÖLLER'S CHAMELEON
Kinyongia uthmoelleri 18–23cm

A fairly large chameleon with short, *scaly horn* on snout tip and *elevated ridges* along each side of snout. These, and tail, are smaller in females. *Crests absent* from body and belly, and *tail is longer* than body. Colour varies, but usually consists of irregular bars of green and reddish-brown on body, with thin *white lateral line*. Endemic to northern Tanzania, where it occurs in forest on rim of Ngorogoro Crater and adjacent hills, and also in South Pare Mountains. Lives in montane forest patches, often high (up to 10m) in trees. A clutch of 10 eggs has been noted.

TWO-STRIPED CHAMELEON
Trioceros bitaeniatus 12–16cm

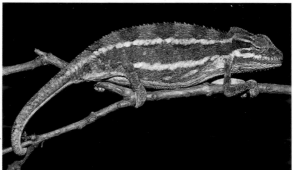

Colin Tilbury

A smallish, brown-grey chameleon, with *two distinctive white side stripes*. Some individuals develop 4–5 bars on flanks. Males lack horns and casque is only slightly pronounced. Has *1–2 rows of enlarged tubercles* on flanks and *small gular crest extends onto belly*. Found in highland regions of Kenya and Uganda, just enters northern Tanzania, and populations also occur in Ethiopia. Prefers small trees and grassland. Gives birth to 6–25 minute tan-brown young.

VON HÖHNEL'S CHAMELEON
Trioceros hoehnelii 15–20cm

Colin Tilbury

A small chameleon with *massive helmet-like casque* and *beard of elongate gular scales*. Body scales are mixed, with small granules and very large tubercles. Body coloration varies from green-grey to yellow-brown, usually with two faint, thin, pale side stripes. Restricted mainly to high grasslands in Kenya, just reaching Mt Elgon region. Inhabits small shrubs and tall grass tussocks, and gives birth to 7–18 young.

JACKSON'S THREE-HORNED CHAMELEON
Trioceros jacksonii 20–38cm

The classic horned chameleon of East Africa. Males (above) bear three long rhinoceros-like horns that may be longer than head. If present

in females (right), horns much smaller. *Vertebral crest well developed, while gular crest is absent.* Colour variable, but usually shades of green, often with enlarged geometric blotches on flanks. Inhabits woodland highlands around Nairobi and Mt Meru. Males territorial and fight using their horns. Up to 28 live young born in wet season.

TUBERCLE-NOSED CHAMELEON
Trioceros tempeli 15–24cm

Medium-sized chameleon that looks like a lichen-covered log. Has *small ear flaps, swollen snout, double gular crest of spiny scales,* and

scattered *large tubercular scales on flanks.* Coloration varied – often mottled green, but can also be blotched greys, blacks and creams. Found in high-altitude forest in southern Tanzania. Up to 28 young are born in the wet season, and are perfectly camouflaged.

MELLER'S GIANT ONE-HORNED CHAMELEON
Trioceros melleri 30–59cm

The largest African chameleon. Its bulk is enhanced by stiff, *blade-like nasal horn, large ear flaps,* and *scalloped dorsal crest* that also extends along tail. Body leaf green, usually with *3–4 irregular yellow bands.* Females lack nasal horn. Lives in woodland and riverine forest, climbing up to 10m high and feeding on large insects. Lays very large clutches of up to 90 eggs in hole dug in sun-warmed soil.

MT KENYA DWARF CHAMELEON
Trioceros schubotzi 10–15cm

This small montane chameleon has *well-developed vertebral crest.* Gular crest is weak and does not extend onto belly. Body bright green, usually with broken white line high on flanks and large white spots beneath. Darker, inverted, triangular patches border spine. Lives in stunted scrub high on Mt Kenya. Gives birth to 7–10 live young.

PYGMY GRASS CHAMELEON
Rieppeleon kerstenii 5–10cm

A *minute* chameleon that crawls among leaf litter, protected by superb camouflage and slow movements. Has *unadorned snout* and *short tail, less than half body length*. Claws of hand have *strong secondary cusp*. Colour varies from rust-brown to yellow-grey, often with vague stripes. Found in coastal thicket and woodland, often sheltering in holes and termite burrows. Lays small clutch of up to 10 eggs in soft soil.

PYGMY CHAMELEON
Rieppeleon brachyurus 4–6cm

Another minute terrestrial leaf chameleon. *Snout short and unadorned* and *tail very short, less than ¼ body length*. Claws of hand have *faint secondary cusp*. Coloration *resembles dead leaf*, usually light brown to grey, sometimes with darker belly. Locally restricted to eastern Tanzania, and elsewhere to Malawi and northern Mozambique. Walks among leaf litter, feeding on small insects and ants, and climbs up onto small twigs to sleep. When held in the hand, often gives eerie buzzing vibration. Lays up to 14 minute eggs.

BEARDED PYGMY CHAMELEON
Rieppeleon brevicaudatus 5–9cm

Another minute terrestrial leaf chameleon named for its short *tail, about ⅓ body length*. Claws of hand have *faint secondary cusp*. Scales along backbone form *low crenulated crest*. Coloration usually light brown to grey with faint stripes, resembling dead leaf. Inhabits lowland forest patches in eastern Tanzania and the Shimba Hills, Kenya. When held in the hand, often gives eerie buzzing vibration. Lays about five, but up to 10, minute eggs.

NGURU SPINY PYGMY CHAMELEON
Rhampholeon acuminatus 5–8cm

A small arboreal leaf chameleon with *elevated casque* on back of head, and *flat horn* on snout in both sexes. Has *scattered enlarged spines* along *backbone* and *above eye*. Tail much *shorter than body* and *non-prehensile*. Claws of the hand *bicuspid*. Coloration shades of green to light brown. Endemic to Nguru Mountains, Tanzania, and inhabits montane forest, often 3–4m off ground. Lays small clutch of about 2–4 minute eggs.

USAMBARA SPINY PYGMY CHAMELEON
Rhampholeon spinosus 5–9cm

Small, with long snout that has *spiny, laterally flattened, droopy horn* at tip in both sexes. Has *scattered enlarged spines* along backbone and on throat. *Prehensile tail almost as long* as body. *Simple claws.* Body colour a mixture of light brown, grey and light blue, resembling lichen-covered twig. Endemic to forests above 700m on East and West Usambara mountains, climbing on low bushes. Lays small clutch of four small eggs, perhaps twice a year.

ULUGURU PYGMY CHAMELEON
Rhampholeon uluguruensis 4–5cm

A minute terrestrial leaf chameleon restricted to Uluguru and adjacent mountains of east-central Tanzania. Has *small, soft proboscis* on tip of snout, and *very short tail*, about $\frac{1}{3}$ *body length*. Claws of hand have *distinct secondary cusp*. Coloration resembles that of dead leaf, in mottled browns, often with two *dark oblique lines*, like leaf veins. Walks among leaf litter, feeding on small insects and ants. Lays 3–4 minute eggs.

PARE PYGMY CHAMELEON
Rhampholeon viridis 6–9cm

A small, plain, arboreal leaf chameleon that lacks *enlarged spinose scales* on head or body, or horn on snout tip. *Tail relatively long* and *almost half* body length. Claws of hand are *simple.* Body green (emerald green in breeding males), often with two darker, *orange oblique* lines on flanks. Endemic to Afromontane forest patches in Pare Mountains, Tanzania. Climbs in low shrubs, and when held in the hand often gives eerie buzzing vibration. Lays 4–8 small eggs.

EAST USAMBARA PYGMY CHAMELEON
Rhampholeon temporalis 5–8cm

A very small terrestrial leaf chameleon with *short, soft horn* on tip of snout, but lacking enlarged spiny scales on head and body. *Tail relatively long, about 1/3 body length.* Claws of hand are *simple.* Coloration dull grey-green in male.

Female resembles dead leaf. Single dark stripe may run obliquely across flank. Restricted to Eastern Usambara Mountains. Walks among leaf litter, feeding on small insects and ants, and climbs up onto small twigs to sleep. Lies still when held in the hand, and not known to make buzzing vibrations. Lays eggs but little else known of its biology.

GECKOS Family Gekkonidae

Widely distributed throughout the world. All species have *immovable eyelids*, and use the tongue to wipe eyes clean. Many have *complicated clawed toes*. Expanded toe-tips have specialised scales covered in minute hairs (scansors), which facilitate climbing of seemingly smooth surfaces. Mostly nocturnal, with large eyes and complicated pupils that constrict to 2–3 small holes, but open fully at night. All lay 1–2 hard-shelled eggs.

USAMBARA FOREST GECKO
Cnemaspis africana 5–10cm

Small, with *flattened body*. Feet have *long, clawed digits that lack scansors*. Has *9–12 pores* in front of anus. Body olive green mottled with brown, with pale line connecting diamond-shaped spots along backbone. Throat white, rest of belly yellowish. Nocturnal, arboreal and shelters among tree buttresses and rock cracks in woodland and hill forest. Lays two hard-shelled eggs in rotting logs or under bark.

ULUGURU LEAF-TOED GECKO
Urocotyledon wolterstorffi 5–10cm

Small, flat-bodied forest gecko. Has short toes with tips *flared into leaf-shaped scansors*.

Paul Freed

Another scansor occurs on tail tip and aids climbing. Eyes orange with vertical pupils, and body mottled grey and cream. Juveniles have reddish tails, but adults have grey tails with 5–6 broad dark bands. Nocturnal, living in montane forests of Usambara and Uluguru mountains. Shelters in hollow trunks and under bark. Lays two hard-shelled eggs at regular intervals.

EAST AFRICAN HOUSE GECKO
Hemidactylus angulatus 10–15cm

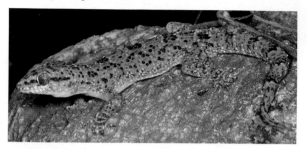

A small terrestrial gecko with *flared toe-tips* that have *paired scansors* and *large retractile claws*. Skin covered with *14–25 rows of strongly keeled tubercles*. Body light brown with rows of enlarged red-brown blotches across back. Dark line runs from nostril through eye to neck. Widespread in northern regions in diverse habitats. Nocturnal, emerging after dark to feed on insects. Adapts well to urban areas and can be found in houses. Lays pair of hard-shelled eggs at regular intervals.

BARBOUR'S GECKO
Hemidactylus barbouri 5–9cm

A small, gentle, terrestrial gecko with *flared toe-tips* that have *paired scansors*. Males have only *16–23 pores* across thighs, and *mainly smooth back scales*. Hides among coral rag exposures in coastal forest, from northern Tanzania just into southern Kenya. Considered relatively rare, but may simply be difficult to find. Body covered in *scattered, enlarged, feebly keeled tubercles*. Grey-ochre body may have scattered pale spots or diffuse crossbars, and vague dark line may pass through eye onto back of head. Lips and belly white-cream.

TROPICAL HOUSE GECKO
Hemidactylus mabouia 12–15cm

A common species, widespread throughout most of region. Adapts readily to towns and cities and rapidly expanding its range. Although normally found in trees, sheltering under bark or in hollow logs, adapts easily to houses. Has *large, flared toe-tips* that have *paired scansors* and large *retractile claws*. Has *12–18 irregular rows of weakly keeled tubercles* on back and *22–40 pores* beneath thigh. Body pale grey with 4–5 wavy dark crossbars that fade in light. Males territorial and will fight each other. Communal egg sites may contain up to 60 eggs.

FLAT-HEADED GECKO
Hemidactylus platycephalus 12–18cm

A very large arboreal gecko often found together with, and difficult to distinguish from, Tropical House Gecko. Both adapt well to towns and are found together on house walls feeding on moths and other insects attracted to lights. Bigger than Tropical House Gecko and usually has only *10–14 irregular rows of weakly keeled tubercles* on back, but *more pores beneath thigh (45–57)*. Body pale grey with 4–5 wavy dark crossbars that fade in light. Inhabits mainly wooded areas, and less likely to be found in rock cracks than is Tropical House Gecko. As befits its size, is more aggressive and will attack, even eat, other small geckos. Also calls in evening, giving long series of up to 15 sharp clicks. Often favours baobab trees. Lays pairs of eggs under bark or in tree hollows.

NYIKA GECKO
Hemidactylus squamulatus 5–9cm

A widespread terrestrial species that lives beneath rocks and in burrows in savanna and semi-arid scrub. Body *lacks enlarged tubercles*, but is covered in *overlapping scales* with *10–16 rows of large, keeled scales*. Coloration light brown, flecked with dark brown and white. Dark line runs from nostril, through eye to back of head. Completely nocturnal and mainly solitary, emerging to hunt termites and other insects, and usually only discovered when sheltering under a stone or dead log. Lays pair of hard-shelled eggs at regular intervals.

GREEN DAY GECKO
Phelsuma dubia 10–15cm

A small arboreal gecko of coastal region. Lives on bushes and tree trunks. Body covered in *small granular scales without enlarged tubercles*. Males have *arched row of 19–29 pores* from thigh to thigh. *Toes flared at tips*, except for greatly reduced *vestigial inner toes*. Body *dull green*, flecked with bright green scale tips and with *irregular tan-orange blotches on back*. Eye has sky blue ring. Completely diurnal. Males usually solitary and may fight other males. Green day geckos feed on insects, but also lap tree sap and nectar from flowers. Lays pair of hard-shelled eggs between palm fronds or under bark, sometimes in communal sites.

MONTANE DWARF DAY GECKO
Lygodactylus angularis 4–8cm

All dwarf day geckos *diurnal* and have feet with *paired oblique scansors*, large *retractile claws*, and *vestigial inner toes*. This small, dark species has 7–10 preanal pores. Throat *yellow with 'V'-shaped dark lines*. Back mottled in browns and greys, with scattered black and white spots. Lives in mountain forests and on forest edges, running up tree trunks in clearings. Large communal egg sites may contain hundreds of old and new eggs.

USAMBARA DWARF DAY GECKO
Lygodactylus gravis 5–9cm

A large dwarf gecko endemic to Usambara and South Para mountains. Back grey-brown with scattered black and white spots, sometimes with two pale lateral bands. Below belly, inner thighs and tail yellow-orange.

Eyes red and throat *cream to light yellow with scattered dark flecks*. Forages on tree trunks and on ground in leaf piles, particularly next to nests of ants, which form main diet, along with other insects. Lays pair of hard-shelled eggs at regular intervals and may use communal nesting sites. Vulnerable.

CAPE DWARF DAY GECKO
Lygodactylus capensis 6–7cm

A delightful dwarf gecko often seen running on trees or garden walls. Has *rudimentary inner toe*, while other toes have *dilated tips, with large retractile claws* and *paired oblique scansors*. Grey-brown body has dark streak from snout to shoulder. Throat usually stippled with grey, belly cream. Feeds almost exclusively on ants and termites. Breeding is continuous and communal egg sites common. If disturbed, may freeze or quickly run for cover.

TURQUOISE DWARF DAY GECKO
Lygodactylus williamsi 4–5cm

One of region's most beautiful geckos. Endemic to small patches of screw pine (*Pandanus rabiensis*) habitat in central Tanzania, and considered Critically Endangered due to unsustainable collecting for pet trade. Has *rudimentary inner toe*, while other toes have dilated tips, with large retractile claws and paired oblique scansors. Adult males bright turquoise above, yellow-orange on belly, with black throat marks. Female greenish-bronze. Lives almost exclusively on leaves of large screw pines. Females breed continuously and lay clutches of two eggs between leaf fronds.

YELLOW-HEADED DWARF DAY GECKO
Lygodactylus picturatus 4–9cm

One of the most common conspicuous dwarf day geckos in region. Males develop *bright yellow heads* and grey bodies, while females remain less conspicuously coloured. Grey blotches and lines on head and body more evident when gecko is cool or unaroused. Adapts well to villages and shambas, climbing onto tree trunks and hut walls in search of insects. Also laps flower nectar. When disturbed, moves around tree trunk to shield itself from danger. Lays pair of hard-shelled eggs at regular intervals.

WHITE-HEADED DWARF DAY GECKO
Lygodactylus mombasicus 4–9cm

Similar to previous species. Distinguished by *white head with black markings*, and grey body with *two dark dorsolateral stripes* that extend from neck onto forebody. Restricted to coastal regions, where it is common in small trees and around villages. Biology and reproduction similar to those of its yellow-headed cousin.

TUBERCULATE THICK-TOED GECKO
Elasmodactylus tuberculosus 12–17cm

A stocky gecko with large head, rounded snout and big eyes. Skin rough, composed of rows of *enlarged, keeled scales* separated by *small granular scales*. Fat toes have *swollen scansors* with 10–12 lamellae. Males have 6–8 pores in front of anus. Body dark brown, with series of paler blotches along backbone. Juveniles have six pairs of dark blotches bordering spine. Restricted to central Tanzania extending across to Zambia. Partially terrestrial, often found under dead trees and in houses. Lays clutches of hard-shelled eggs throughout wet season.

TURNER'S THICK-TOED GECKO
Chondrodactylus turneri 13–18cm

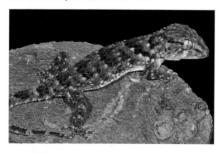

A large, stout gecko with rough skin composed of rows of *enlarged, keeled scales* separated by small granular scales. Fat toes have *swollen scansors* with *10–12 lamellae.*

Males have *no pores* in front of anus. Has broad, muscular head and can give powerful bite. Prominent *black-edged white spots cover back.* Shelters in rock cracks or under dead trees, and usually found singly or in pairs. Found in savanna in scattered populations in southeastern Tanzania, Kenya–Tanzania border region and possibly Rwanda. Lays multiple clutches of paired eggs during wet season, which hatch in 60–80 days.

BANDED VELVET GECKO
Homopholis fasciata 13–16cm

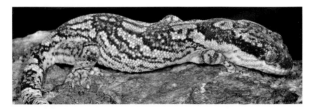

Fairly large, stout gecko with short head, ringed with black neck band. *Snout round* with *white lips*. Skin *smooth*. Toes have 8–12 '*V*'-shaped scansors and a *retractile claw*. Males have only *two pores* in front of anus. Body fat with *oblique dark and light bands*. Fat tail also *banded* in light and dark rings. Shelters in holes in big trees, particularly baobabs. Found in scattered localities in savanna of Tanzania and Kenya. Multiple clutches of two eggs are laid.

LIDDED GECKOS Family Eublepharidae

Terrestrial geckos that have movable eyelids and typical clawed toes that lack scansors. All are nocturnal.

EAST AFRICAN LIDDED GECKO
Holodactylus africanus 8–11cm

Left: *Adult* Right: *Adult and juvenile*

Stout, with broad head, prominent eyes and *movable eyelids*. Body covered in *uniform small granules*. *Tail short*, either swollen or slender. Body reddish-brown to chestnut with pale and dark chevron bands. Pale line runs along backbone from back of head to tail tip. Juveniles more brightly banded in wavy chocolate and cream bands. Restricted to patches of Somali–Maasai bushland in northern Tanzania and Kenya and north to Somalia. Slow-moving and terrestrial. Emerges from burrow at night to forage for termites and other insects. Lays two hard-shelled eggs in sandy soil.

WORM LIZARDS Family Amphisbaenidae

Legless burrowing lizards that look like earthworms owing to elongate bodies covered in rings of flat, rectangular scales. Lack external eyes. Heads shaped to facilitate pushing through soil.

MPWAPWA WEDGE-SNOUTED WORM LIZARD
Geocalamus modestus 20–28cm

A worm lizard with a *laterally compressed snout*. Pectoral shields on the throat *slightly enlarged* and form *arrow-shaped series*. Head and body violet to grey-brown, paler below. Apparently rare and poorly known, but probably overlooked. Found in Dodoma region, but probably more widespread in central arid parts of Tanzania.

CROCODILES Order Crocodilia

Aquatic remnants of the dinosaurs' rule. The 28 species are distributed throughout the tropics, with the African radiation now comprising seven species. Many now endangered.

EASTERN DWARF CROCODILE
Osteolaemus osborni 100–130cm

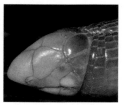

Small, with a short, very broad-snouted head and prominent eyes. Tail has fat base, narrows rapidly, and is almost as long as body. Appears dark, being mottled in black, brown and olive. Juveniles more brightly patterned. Inhabits eastern Congo rainforest, just reaching western Uganda. Shelters in small forest streams and swamps. Active mainly at night, feeding on frogs and crabs on forest floor. Lays up to 20 eggs in mounds of rotting vegetation. Widely hunted for meat. Endangered.

NILE CROCODILE
Crocodylus niloticus 250–550cm

Second-largest living crocodile, may exceed 1,000kg. Males larger than females. Eye has extra eyelid that sweeps away dirt. Eyes and valved nostrils placed high on head. Hindfeet webbed and long tail has two raised dorsal keels. Adults dull

olive with yellow or cream belly. Hatchlings brighter, with irregular black markings and straw-yellow belly. Female digs hole on sandbank and lays 16–80 hard-shelled oval eggs. Incubated by the sun, they hatch in about 85 days. Incubation temperature determines sex of hatchlings. Both parents protect nest during development, assist hatchlings from nest and carry them to water. Growth is slow, and maturity occurs in 12–15 years, at 2–3m. Youngsters live in marshes and backwaters, feeding mainly on insects and frogs. Adults move to more open water holes or river pools, where they ambush mammals and birds. Fish, particularly catfish, form main diet in many populations (e.g. Lake Turkana).

SLENDER-SNOUTED CROCODILE
Mecistops cataphractus 150–300cm

A *medium-sized* crocodile distinguished by its *long, narrow snout*. Body dark olive in adults. Juveniles more vividly marked, with lighter grey-green body and irregular dark bars and blotches. Found throughout Central Africa, but locally restricted to Lake Tanganyika. More aquatic than other species, and eats fish almost exclusively. Lays up to 30 eggs in mound of rotting vegetation. Non-aggressive, except when defending nest site.

CHELONIANS   Order Testudines

A very ancient group whose members are recognised by their protective shells, consisting of horny outer scutes covering an inner bony layer. Outer scutes may have been lost in some aquatic species. Internal bony shell may also be reduced in size.

SIDE-NECKED TERRAPINS   Family Pelomedusidae

Primitive chelonians that retract neck sideways when head is withdrawn. Restricted to the southern continents.

NEUMANN'S MARSH TERRAPIN
Pelomedusa neumanni 17–20 cm

Tomas Mazuch

A small terrapin with a *very flat, hard shell that lacks a hinge*. The neck is long and muscular, there is *one large, undivided temporal head scale* and there are *two soft tentacles on the chin*. Coloration variable above and below, from horn-coloured to almost black. They inhabit shallow pans and marshes, and aestivate underground during the dry season. The diet is varied, including plant material and aquatic insects. Presently known from Kakamega to Manyatta and Lake Victoria, but possibly more widely distributed. Mating occurs in water, and up to 20 soft-shelled eggs are laid in a sandbank.

SERRATED HINGED TERRAPIN
Pelusios sinuatus 30–55cm

Large freshwater terrapin with a *hard, domed* shell that protects it from crocodiles. Front of *plastron hinged* and closes to protect head. Rear of shell *serrated*. Juveniles have *keels along backbone* that disappear in large

adults. Shell black, except for *angular-edged yellow blotch in centre* of belly. The most common terrapin in region, particularly in larger rivers and lakes (not Lake Victoria). Often spotted basking on floating logs. Lays 7–30 soft-shelled eggs. Eats water snails, soft waterweed and insects, and often seen pulling ticks off large mammals.

YELLOW-BELLIED HINGED TERRAPIN
Pelusios castanoides 18–23cm

A medium-sized terrapin with an *elongate smooth shell* and *small plastral hinge*. Shell is olive, blackish-brown or yellowish, with *yellow plastron*, usually with faint black markings. Skin on neck and limbs *yellow*.

Frequents shallow water in still lakes and swamps at low altitudes. As the dry season approaches, buries itself in mud, re-emerging with the rains. Many specimens have scarred shells from fires during dry season. Feeds on aquatic insects, frogs, freshwater snails and floating vegetation. Lays clutches of up to 25 soft-shelled eggs.

WILLIAM'S HINGED TERRAPIN
Pelusios williamsi 20–25cm

Colin Tilbury

A large hinged terrapin with an *oval shell* and *slight vertebral keel* that is more prominent in juveniles. Shell usually *black to dark brown* above and below, but sometimes with paler plastron and yellow midline. Juvenile has black-and-yellow vermiculated head, which darkens with age. Inhabits rivers and swamps draining into Lake Victoria, but is shy and rarely observed basking. Lays clutches of soft-shelled eggs.

FOREST HINGED TERRAPIN
Pelusios gabonensis 18–25cm

Marius Burger

One of the few hinged terrapins found in lowland forest. Shell *flattened with keel along backbone*. Head has *broad, dark Y-shape between eyes*, while shell colour varies from straw yellow to dark brown, usually with dark stripe along backbone and a black belly. Inhabits forest pools and swamps, but is shy and only rarely basks on river banks and floating logs. Although eaten by forest people, like many terrapins gives off pungent smell. Just enters forest patches in extreme western Uganda. Lays clutches of up to 12 soft-shelled eggs.

PAN HINGED TERRAPIN
Pelusios subniger 14–20cm

A small hinged terrapin with a *smooth, rounded shell* that is uniform brown with yellow and brown junction to upper and lower shell. Dark plastron shields have pale yellow centres, and skin on neck and limbs is grey or black. Found in pans and temporary water bodies, and buries itself underground during dry season. Small frogs and invertebrates form main diet. Lays up to eight eggs. May discharge cloacal contents in self-defence.

LAND TORTOISES Family Testudinidae

**Advanced chelonians: can withdraw *head backwards into shell*.
Hindfeet elephant-like. They walk on tips of armoured forefeet.
Terrestrial, mainly vegetarian and lay hard-shelled eggs.**

LEOPARD TORTOISE
Stigmochelys pardalis 30–55cm

Left: *Adult* Right. *Egg and hatchling*

Largest East African tortoise, distinguished by its *high, domed shell* and *lack of nuchal scute* at front of carapace. *Gulars divided, carapace lacks hinge, and there are 2–3 buttock tubercles* on each side. Hatchlings bright yellow, each scute having irregular central blotch. Adults become darker and heavily blotched or streaked. Very old tortoises may be almost uniform dark grey. Males have hollow bellies and longer tails. Occurs in central part of region in moist and arid savanna, not forest. Lays up to three clutches of 6–15 eggs at intervals through wet season. These may take 12 months to hatch.

PANCAKE TORTOISE
Malacochersus tornieri 12–18cm

Small with *very flat and flexible shell* that *lacks hinge*. Head rounded and beak has *2–3 cusps*. Males have longer and thicker tails. Shell brown, each scute with pale centre and radiating pattern of pale lines. Juvenile's shell vividly marked with orange and black. Endemic to East Africa, in dry savanna with scattered rock outcrops. Shelters in rock cracks, often in small groups, and climbs easily. Eats herbs, succulents and beetles. Lays single large egg at start of rainy season. Endangered by over-collection for pet trade.

EASTERN HINGED TORTOISE
Kinixys zombensis 17–22cm

Medium-sized with characteristic *hinge at rear of carapace* that closes when tortoise is disturbed, protecting hindfeet and tail. Carapace slightly domed, scutes usually have *radiating pattern of dark bands*, and it *lacks buttock tubercles*. Restricted to regions of moist savanna and coastal thicket. Emerges in early morning and evening to feed on fruits and soft plants. Also eats snails and millipedes. Lays 2–7 eggs (up to 10 in exceptional cases).

FOREST HINGED TORTOISE
Kinixys erosa 20–35cm

Fairly large with characteristic *rear hinge in carapace* that closes to protect hindfeet and tail. Carapace *slightly domed* and rich red-brown, with black edges to scales, and cream lateral line. Marginal

scales *heavily serrated* and *recurved upwards*. Adult males have concave plastron. Restricted to Congo forest, just reaching western Uganda. Secretive and now rare, has been over-hunted for bush meat. Emerges in early morning, from shelter in hollow logs or among tree roots, to feed on fruits, mushrooms, snails and millipedes. Lays about 4–8 large eggs.

SOFT-SHELLED TERRAPINS Family Trionychidae

Terrapins with only three toes on each foot and soft, leathery shells. Have bony layer beneath skin. Restricted mainly to Asia and North America, with only five species in Africa.

ZAMBEZI SOFT-SHELLED TERRAPIN
Cycloderma frenatum 35–50cm

Large (up to 14kg) with *very long neck* and '*snorkel-like*' *nose*. Hindlimbs protected by *flexible skin flaps* when withdrawn into *soft shell*. Uses strong forelimbs to dig in soft mud for snails and mussels, using powerful jaws to crush them. Often shuffles into sand, leaving only head exposed. Lays 15–22 hard-shelled eggs on sandbank, often near crocodile nest. Restricted to river systems in southern Tanzania. Increasingly threatened by illegal collecting for Asian food market.

NILE SOFT-SHELLED TERRAPIN
Trionyx triunguis 40–120cm

The *largest African terrapin*: females may reach 120cm and over 60kg. Males smaller (up to 65cm). Plastron *lacks flexible flaps* over hindlimbs. Head *elongate and flattened*. *Snorkel-like snout.* Forelimbs have *three sharp-edged, crescent-shaped skin folds.* Shell olive to dark reddish-brown in old adults, but usually with light-centred dark spots, often bordered with yellow, in juveniles. Restricted to the Nile drainage below Murchison Falls, and to Lakes Turkana and Albert. Ambushes fish, crabs and frogs from under sand or water vegetation.

SEA TURTLES Superfamily Chelonioidea

Sea turtles have front feet modified into flippers and *are unable to withdraw head or feet into their shells*. Found throughout tropical seas, they return to sandy beaches to lay soft-shelled eggs. Coastal waters in the region are home to five species, but only two are common.

GREEN TURTLE
Chelonia mydas 98–120cm

Among the commonest sea turtles in region. Enters shallow estuaries to feed on sea grasses and other forms of marine life such as jellyfish. *Shell hard, smooth, with non-overlapping scutes.* Has *12 marginals* on each side, which are smooth in adults. Front flippers each have *single claw*. Females usually darker in colour than males and grow slowly, taking 10–15 years to mature. Threatened by pollution and slaughtered for meat and eggs. Nest throughout the year, but few large aggregations remain, except on protected islands.

HAWKSBILL TURTLE
Eretomochelys imbricata 60–90cm

A small sea turtle with a *hard, oval shell* that has *thick overlapping scutes*. Has *12 marginals* on each side, of which those at the back are

markedly serrated. Front flippers each have *two claws*. Feeds on coral and urchins prised from sea floor with *hooked beak*. Many millions have been killed for their shells and used to make 'tortoiseshell' jewellery. Nest only on small coral islands, the nearest being northeastern Madagascar and Mauritius.

OLIVE RIDLEY TURTLE
Lepidochelys olivacea 50–73cm

The smallest sea turtle, may weigh 45kg, and with *heart-shaped, smooth, flat-topped shell.* Has numerous scutes, with *five vertebrals* and *5–9 costals* on each side. *Each limb has two claws.* Shell dark to light olive green above, with

pale yellow, almost white plastron. Prefers shallow coastal waters and major estuaries, and feeds, sometimes at considerable depths, on bottom-living crustaceans, particularly prawns and shrimps. Not known to nest on African beaches any more. Elsewhere in western Indian Ocean nests in Oman and on east coast of India.

LEATHERBACK TURTLE
Dermochelys coriacea 130–170cm

Does not breed in region, and only a few recorded in coastal waters. The *largest sea turtle*, easily recognised by *pliable rubbery shell*, which has *12 prominent ridges.* Young are *blue-grey* with *long flippers.* A specialist feeder on jellyfish, travelling the ocean currents in search of its prey. May dive to depths of over 300m, spending up to 37 minutes underwater. Growth very rapid and sexual maturity may be reached in only 5–7 years. Females lay up to 170 eggs at a time, and visit nesting beaches up to five times per breeding season.

AMPHIBIANS   Class Amphibia

FROGS   Order Anura

AFRICAN CLAWED FROGS   Family Pipidae

Primitive aquatic amphibians found only in Africa and South America. Have no tongue, eye has circular pupil and there is lateral-line sense organ along side of body. Some hind toes clawed.

NORTHERN CLAWED FROG
Xenopus borealis 50–90mm

A *flat-bodied, smooth-skinned* frog with *unwebbed, long-toed forelimbs* and *strongly webbed hindlimbs*, on which *three toes are black-clawed*. Has *short tentacle* below each eye. Back colour grey-brown with many irregular dark markings, dense

on hindlimbs and flanks. Pale belly, usually lightly spotted. Almost fully aquatic and prefers pools or slow-moving water. Eats aquatic insects, small fish and even own tadpoles. The latter have *transparent bodies, filamentous tails and large tentacles*, and hang head-down filtering plankton. Found in highlands of Kenya and just into Tanzania.

SNOUT-BURROWERS   Family Hemisotidae

Burrowing frogs, restricted to Africa. Snout hard and pointed, pupils vertical. The only frogs to burrow headfirst into damp soil.

MARBLED SNOUT-BURROWER
Hemisus marmoratus 30–55mm

Small burrowing frog with *smooth skin*. Has *fold between and above eyes*. Forelimbs short and muscular to aid digging. *Hindlimbs only slightly*

webbed. Back yellowish-brown and very mottled. Emerges to feed on insects at night. Lays 150–200 eggs in a tunnel in wet soil leading from standing water. Female stays with clutch and large tadpoles emerge after eight days and wiggle to water. Male gives repeated buzzing call from mud near water.

TOADS Family Bufonidae

Stocky amphibians with dry, warty skin. They usually have a pair of parotid glands on sides of neck that secrete toxins. Toads move about with laboured hops and are common around homes.

GARMAN'S TOAD
Amietophrynus garmani 80–115mm

A large toad with conspicuous parotid glands and *black-tipped warts* on tan or reddish-brown back. Also has *reddish inner thighs*, but top of head plain, with *dark broken bar behind eyes*. Inhabits savanna or secondary woodland, and breeds in shallow ponds and still water. Females lay double strings of up to 20,000 eggs. Males give loud *'kwaak' call* in small choruses from water's edge.

GUTTURAL TOAD
Amietophrynus gutturalis 70–120mm

A common, very large and noisy toad, with *very large, smooth parotid glands* that produce a

powerful toxin. Light brown back has paired dark-edged blotches, and often a thin pale line along spine. Top of head has characteristic *pale cross* running between eyes. *Inner thighs and groin red*. Widespread in savanna and coastal habitats and common around houses. Large numbers congregate at shallow pools when breeding. Males give *slow 'snoring' call*, and females lay long strings of up to 25,000 eggs.

KISOLO TOAD
Amietophrynus kisoloensis 50–87mm

A large, stocky toad with *pointed snout*. *Parotid glands large* and male's back smooth, while that of female is warty. More aquatic than many toads, with *slender, extensively webbed digits*. Body boldly marked, with a greenish hue, dark purple-brown blotches and, often, thin light line along spine. Breeding males turn *bright yellow* and give slow snore, while hiding in vegetation beside shallow streams. Tadpoles brownish in colour. Restricted to cool, moist mountain forests of Kenya, extending south to northern Zambia and Malawi.

FLAT-BACKED TOAD
Amietophrynus maculatus 40–80mm

A small savanna toad with *flattened parotid glands* usually covered with *black-tipped spines*. Has *row of white tubercles under each forearm*, and toes *only slightly webbed*. Body mottled in light tan and cream, with *light cross between eyes*. Belly cream speckled with grey, and breeding males have blackish throat. Breeds in shallow pans in open savanna, and female lays double string of up to 8,000 eggs. Male's call a brief trill, given from vegetation.

SUBDESERT TOAD
Amietophrynus xeros 50–95mm

Steve Spawls

A large toad with *narrow parotid glands* that start close to eye. Body pale brown with six pairs of darker blotches along back, and scattered dark spinules. Inner surface of thighs *bright red*. Belly cream, and breeding males have dark throats.

Prefers arid savanna, from Algeria and Senegal east to Ethiopia, and through much of Kenya, with scattered records in Tanzania. Breeds in shallow pools in riverbeds and in oases, and survives dry season by burrowing deep down in mudbanks.

RED TOAD
Schismaderma carens 60–90mm

A large toad with *elongate glandular ridge* that extends from neck and along flanks. *Tympanum as large as eye.* Back *rust red*, with *paired blotches* on shoulders and lower back, demarcated from pale flanks by dark border.

Belly pale with grey speckles. Terrestrial, emerging after dark from burrow or shelter under a stone or log to forage for crickets and slugs. Breeds in shallow ponds and marshy pools. Just enters southern Kenya. Lays up to 20,000 eggs in double strings. Males give low, booming call while floating in water.

TAITA DWARF TOAD
Mertensophryne taitanus 20–33mm

Underside

A small toad of the woodland floor, with *slender body* and *thin legs*. Has small *visible tympanum* and *flattened parotid glands* that are continuous with eyelid. Toes *slightly webbed*. In males back light gray or tan with three pairs of dark marks. Females have uniform pale back and darker flanks. Belly white, with characteristic trident pattern. Found from Taita Hills, Kenya, through woodlands of northern Tanzania to northwest Zambia, and south along Tanzania coast. Lays small clutch of up to 150 eggs in strings in small muddy pools. The tadpole lacks the fleshy circular crown seen in Woodland Dwarf Toad tadpole.

WOODLAND DWARF TOAD
Mertensophryne micranotis 15–25mm

Mark-Oliver Rödel

A small toad of the forest floor that has dark brown back, pale patches on shoulder and pelvic regions, and mottled belly. *Tympanum not visible* and *parotid glands are indistinct. Outer toe reduced in size.* Males have rough thumbs to aid in clasping females during mating. Lays small clutches of 8–12 eggs in small puddles of water in fallen trees or even old snail and coconut shells. Tadpole has fleshy circular crown that assists with breathing in small stagnant pools.

KIHANSI SPRAY TOAD
Nectophrynoides asperginis 10–20mm

A very small, smooth-skinned toad that *lacks parotid glands and tympanum*. Toes slightly webbed. Body yellow to mustard-yellow, with paired dark brown lateral bands that are prominent in juveniles but fade in adults. Restricted

to fringe of mossy herbaceous vegetation bordering the Kihansi Waterfall falling from the Udzungwa escarpment, Tanzania. Discovered only in 1998 and Critically Endangered after a hydroelectric scheme destroyed its habitat. Became extinct in the wild, but habitat has been repopulated with captive-bred toads. Breeding occurs all year round. Fertilisation is internal, females giving birth to a few fully formed toadlets.

ROBUST FOREST TOAD
Nectophrynoides viviparus 30–60mm

Small, *smooth-skinned*, with a stocky body. Eardrum *very small*. Parotid glands *very large*, as are *pale-coloured glands on the limbs*, which *contrast* with body. Toes are slightly webbed. Body colour very varied, from grey to pale grey or dark olive, and belly may be

pure black or white. Endemic to Tanzania, but widespread through southern mountains. Shelters among leaf litter and dead logs on forest floor, but may climb up to a metre high in low shrubs. Breeding occurs all year round and fertilisation is internal. Female gives birth to up to 70 toadlets, which she may carry for some time on her back.

SQUEAKERS AND TREE FROGS Family Arthroleptidae

Comprises two very dissimilar groups, restricted to Africa. In squeakers third finger is elongated, especially in males. Their eggs are laid on land in damp leaf litter and develop directly into small froglets, without a free-swimming tadpole phase. Tree frogs were previously allied with reed frogs (Hyperoliidae). They have normal third fingers and free-swimming tadpoles.

COMMON SQUEAKER
Arthroleptis stenodactylus 30–45mm

A stocky litter frog with a broad head and short legs. Body brown, usually with a darker, three-lobed figure and sometimes a light mid-dorsal stripe. *Dark streak* curves from eye to arm. Belly pale, and breeding males have grey mottled throats and an elongated third finger. Widespread in forests and cleared areas from South Africa to Angola. Males give repeated high-pitched chirp while sheltering in leaf litter. Deposits 33–80 yolky eggs directly in moist leaf litter.

DWARF SQUEAKER
Arthroleptis xenodactyloides 13–22mm

A very small leaf litter frog with narrow, wedge-shaped head and short legs. Body coloration highly variable. Usually has dark hourglass figure on brown back, and dark streak from snout tip to ear. Some individuals have plain reddish backs. Toe-tips taper to sharp point, and third finger of males is elongate, edged with small spines to grasp female during mating. Restricted to forests, where it hops in leaf litter. Breeding males give high-pitched chirp. Yolky eggs develop directly into miniature froglets.

EASTERN FOREST SQUEAKER
Arthroleptis adolfifriederici 30–42mm

A stocky leaf litter frog with broad head and short legs. Body mottled brown, usually with marbled sides and sometimes a pale mid-dorsal stripe. *Dark streak* curves from eye to arm. Males smaller (up to 32mm) and

when breeding have grey mottled throats and elongated second and third fingers that grow tooth-like granules. Found in scattered high-rainfall forest patches, from southern Kenya to Malawi and west to the Democratic Republic of the Congo. Males give a series of short, loud chirps from moist leaf litter.

SILVERY TREE FROG
Leptopelis argenteus 30–40mm

A small, stout tree frog that lives in moist savanna and coastal bush. Always has light brown back with inverted dark triangle on neck, and usually has dark eye stripe. *Toes lack enlarged terminal pads* and have

variously striped *reduced webbing*. Brown back (green in Kakamega, Kenya) has *bold 'N'-shaped dark blotch* and there are dark bands running between eyes and from below eye to tympanum. Shelters in hole in the ground, foraging on surface at night for insects. Breeding males give *low 'kwaak' call*, usually from the ground but sometimes in low bushes and trees up to 3m high. Eggs laid in deep holes dug in rain-soaked ground.

BOCAGE'S TREE FROG
Leptopelis bocagii 40–60mm

A savanna species with *hardly any webbing* between toes, which *lack enlarged terminal pads*. Back brown (green in Kakamega, Kenya) with *bold 'V'-shaped*

dark blotch. Has dark bands running between eyes and from below eye to tympanum. Shelters in hole in ground, foraging on surface at night for insects. Breeding male gives *low 'kwaak' call*, usually from ground but sometimes in low bushes and trees up to 3m high. Eggs laid in deep holes dug in rain-soaked ground.

YELLOW-SPOTTED TREE FROG
Leptopelis flavomaculatus 45–70mm

A large tree frog with *broad webbing* between toes, which have *enlarged terminal pads*. Juveniles and some adult males bright green with yellow flecks on back and white heels and elbows. Adult females and some adult males, however, grey-brown with dark brown triangle on back and have dark bands running between eyes and from below eyes to tympanum. Pupil greenish-gold in colour. Call a *long, soft clack* made from dense vegetation up to 4m high, or even from burrow on the ground. Eggs laid among floating vegetation. Inhabits evergreen forest and shelters in hollow tree cavities.

PARKER'S TREE FROG
Leptopelis parkeri 45–70mm

A large forest species in which females grow much larger than males. Slender. Head has *small tympanum*. *Large red eyes* sit high on head and face forward. Toes *bright yellow* and end in *expanded pads*. Inhabits dense high-altitude forest on Eastern Arc Mountains of Tanzania. Breeding males have white throat (yellow in females) and call high in trees, making a quiet buzzing call.

VERMICULATED TREE FROG
Leptopelis vermiculatus 50–85mm

One of Africa's largest tree frogs, with females growing almost twice as big as males. *Toes well webbed* and have *toe-pads*. Juveniles enamel green with black-edged

blotches on flanks. This coloration is retained by some adults. Other adults leaf green, with *brown chevron* on back and irregular white spots on sides. Inhabits low bushes and wild bananas in Eastern Arc Mountain forests. Males give a loud clack call from branches above water.

BURROWING FROGS   Family Microhylidae

Widespread throughout the tropical regions of the world. These small, mainly burrowing frogs run instead of hopping.

RED-BANDED RUBBER FROG
Phrynomantis bifasciatus 40–65mm

A very *smooth-skinned*, brightly striped, burrowing frog. Head *flat and short-snouted*. The thin, almost unwebbed digits of legs have small expanded tips. Body black with pair of *bright red-orange bands*

on flanks that fuse over the hips. Belly grey with white blotches. Skin contains strong toxins. Ants form its main diet. Inhabits savanna in East and South Africa. Males give long, melodious trill from the water's edge. Eggs laid in shallow pools and develop quickly.

KREFFT'S WARTY FROG
Callulina kreffti 30–40mm

An unusual, small, squat, arboreal frog with long limbs. Toes elongated on all feet and have *flattened expanded tips* that aid frog in climbing trees, where it often shelters in

hollow trunks or under dead bark. Body mottled grey-brown and has rough skin. Inhabits summit forests of the Tanzanian Arc Mountains. Lays small clutch of large eggs in a moist hollow, often in dead tree. When disturbed, inflates its body and may exude sticky secretion.

Alan Channing

RAIN FROGS Family Brevicipitidae

Previously included with Microhylids, but now placed in a separate family that is endemic to sub-Saharan Africa.

MOZAMBIQUE RAIN FROG
Breviceps mossambicus 30–40mm

A squat burrowing frog with blunt head and bulbous eyes. *Legs are short and stout.* Body smooth-skinned, usually grey-brown, with dark bar from eye to arm. Belly smooth, mottled in white and grey. The *two outer toes* of the

unwebbed feet are *reduced in size.* Burrows in sandy soils in southern savanna, only emerging after heavy rains. Males call from burrow entrance, giving brief whistling peep. During mating, the small male becomes glued to back of the bigger female. Up to 25 large eggs are laid and undergo direct development.

LONG-TOED RAIN FROG
Probreviceps macrodactyla 30–40mm

This globular, squat frog has warty skin and a *blunt head* with *bulbous yellow eyes.* Body mottled brown with an irregular light yellow side stripe. *Fourth toe very elongate*, but other toes are *unwebbed*

and lack swollen tips. Lives on forest floor, hiding among leaf litter. Lays up to 20 large eggs at end of burrow in moist soil. These develop directly into small froglets that emerge from the egg capsule. Calling males give low-pitched chirp from the ground. This frog inhabits summit forests of the Tanzanian Arc Mountains.

Colin Tilbury

FOAM-NEST FROGS Family Rhacophoridae

Mainly found in Asia, with only three African species. Eye has
horizontal pupil.

PETERS' FOAM-NEST FROG
Chiromantis petersi 50–92mm

These mainly Asian tree
frogs have *sticky, grey
skin*, a large tympanum,
and long legs with
*extensive toe webbing.
Toe-tips are expanded
into pads*. Back grey or
brownish-grey to white,
often with darker
mottling. Inhabit
savanna, including

semi-arid scrublands. Males much smaller than
females, and give quiet, slow, creaking call. Males
and females cling together in amplexus on rock face
or tree branch overhanging water. They rub their
legs together to beat skin secretions into a foam nest
within which eggs are laid. These drop into water
after 3–5 days, where they continue to develop.

GREY TREE FROG
Chiromantis xerampelina 45–90mm

A slender, grey
tree frog that
bleaches almost
completely white
during the day
when basking in
exposed positions.

At night body usually mottled and
resembles tree bark covered in
lichen. *Fingers have expanded tips*
and are arranged in opposable
pairs to clasp vegetation. Both
fingers and toes are webbed.
Breeding males give quiet, squeaky
croak from waterside vegetation. A
single female and several males
construct foam nest on leaves or in
branches overhanging water.
Within this aerial meringue, eggs
develop into small, dark tadpoles, which subsequently fall into the
water below to continue their development.

REED FROGS   Family Hyperoliidae

A large group of diverse frogs restricted to Africa and the Indian Ocean islands. Many have bewildering ranges of colour.

COMMON REED FROGS
Hyperolius viridiflavus-glandicolor complex   30–45mm

Top: glandicolor Above left: mariae *(coastal Kenya)*
Above right: goetzei *(Tanzania)*

A group of small to medium-sized reed frogs of amazing colour variation. Only a few of the more widespread forms are illustrated. Unravelling relationships within the group is currently a scientific nightmare. Skin smooth and sticky, and tips of all digits are expanded into pads. Snout short and hindfeet extensively webbed. Found mainly in savanna, but may enter forest margins. Coloration can bleach when in exposed positions. Form dense breeding aggregations. Males call from vegetation near water, giving series of high-pitched clicks. Clutches of around 2–3,000 small eggs are deposited in open water.

RED-SPOTTED REED FROG
Hyperolius mitchelli 23–32mm

Alan Channing

A forest species, this medium-sized reed frog has *orange to brown* body, often with small dark spots, especially in juveniles. Light snout and dorsolateral stripes are usually black-edged. A light spot present on each heel. Belly orange, sometimes yellow in males. Males give clicking call from vegetation over streams or ponds. A small clutch of eggs is attached to vegetation hanging over water. Tadpoles hatch after 5–6 days and drop into the water below to continue their development.

KIVU REED FROG
Hyperolius kivuensis 25–38mm

Large, with *long snout and slender*, mainly *green body*, with *dark lateral stripe* that passes through *eye to snout*. When it hops, *bright yellow-red* colour of inner limbs is visible. Inhabits open savanna from Angola to Kenya. Breeding males give brief, harsh chirp from high in reeds and flooded grasses. Sticky eggs are laid on vegetation, hatching in just nine days.

SPOTTED REED FROG
Hyperolius substriatus 20–37mm

A large, heavy-bodied, *orange* frog usually with dark-edged pale stripe that edges snout and extends to the neck (sometimes as series of spots to the vent). Other patterns occur including black reticulation or yellow spots outlined in black. A common forest species. Breeding male gives series of short, high-pitched clicks while perched on vegetation around a temporary pool or pond. Eggs attached to leaves overhanging water.

TINKER REED FROG
Hyperolius tuberlinguis 25–35mm

A medium-sized reed frog with *pointed snout* and *uniform coloration*. Body may vary from bright yellow to leaf green or brown, with backward-pointing pale triangle between the eyes.

Inhabits lowland savanna, favouring temporary pools in thick vegetation. Breeding male's call a series of slow clicks, given while sitting in low vegetation. Very sticky all-white eggs laid on vegetation just above the water surface. After early development, tadpoles wriggle free from egg mass and drop into water to continue their growth.

GREATER LEAF-FOLDING FROG
Afrixalus fornasini 25–40mm

A large, slender frog with beautiful striped pattern. Back and tibia have *silvery-white stripes* with *small black-tipped spines*. Silvery bands fade at night. Fingers slightly webbed at the base, and toes webbed. Widespread in savanna and coastal woodland. Male call, a series of rapid clacks followed by a buzz, given from reeds in deep ponds. Eggs laid over water on reed leaves, which are then folded and glued over eggs. These hatch 5–10 days later and tadpoles drop into water to continue their development.

SHORT-LEGGED LEAF-FOLDING FROG
Afrixalus brachycnemis 20–27mm

A minute, *slender frog with thin legs* and adhesive toe pads. Male's body is covered with *small spines,* but these are restricted to head in females. *Body golden* with dark band with light speckling on flanks, and sometimes paired dark lines. Common in eastern savannas and coastal grasslands. Call of breeding male a prolonged buzz, given from flooded grassland.

SENEGAL KASSINA
Kassina senegalensis 25–40mm

A *boldly striped* burrowing frog, in which tips of the *unwebbed digits are expanded into small discs.* Back pale grey with large dark spots and

broad dark vertebral stripe. Spends long periods underground, emerging to eat termites and other insects. Very widely distributed in savannas of sub-Saharan Africa. Calls of breeding males form small choruses from concealed positions in bank-side vegetation. The single abrupt call sounds like cork popping from bottle. Eggs laid in water, attached to submerged vegetation, and the large tadpoles have high tailfins.

RED-LEGGED KASSINA
Kassina maculata 55–68mm

A large, primarily *aquatic* frog in which groin and inner parts of thigh are *bright red.* Back grey-olive covered with large white-edged dark spots, while pale belly usually has dark mottling. Toes and fingertips expanded into *rounded discs.* Widely distributed along East African coastal region. Male's *call sounds like bursting bubble,* given while floating in the water or propped up on aquatic vegetation. Eggs are attached to submerged vegetation, and the large tadpoles are leaf-shaped, sometimes taking as long as 10 months to complete their development.

PUDDLE FROGS Family Phrynobatrachidae

Previously confused with other ranoid frogs, these small frogs are common on forest floors and in moist savanna.

EASTERN PUDDLE FROG
Phrynobatrachus acridoides 20–30mm

A small frog with bumpy skin, usually with *paired chevron-shaped ridges* between shoulders. Toes and fingertips faintly expanded and toes moderately webbed. Body grey and often has green or tan

mid-dorsal stripe or blotch. Belly smooth and pale, usually with yellowish tinge near vent. Male's throat light grey. Generally found in puddles, small streams and on swamp margins. Male gives *continuous, harsh, creaking snore* from shallow water. Eggs float in a single layer just below water surface.

NATAL PUDDLE FROG
Phrynobatrachus natalensis 30–40mm

A small, stout frog that comes in a bewildering array of colours and patterns. Snout pointed. *Toes of short legs lack swollen tips* and usually have ample webbing. Skin lumpy, with broken ridges. Back usually mottled brown, but also often with thin pale stripe or broad red-brown or green band along backbone. Males call at any time during wet weather, giving slow, quiet snore from flooded vegetation. Small eggs float at surface of temporary pond in thin mats. Widespread in savanna throughout region. :

RIDGED FROGS Family Ptychadenidae

Previously confused with other ranoid frogs, these long-hopping frogs are restricted to sub-Saharan Africa, with one species introduced to Madagascar.

MASCARENE RIDGED FROG
Ptychadena mascareniensis 40–60mm

A medium-sized, greenish-brown frog having long legs with only two joints of the fourth toe free of webbing. Skin smooth, except for several rows of parallel ridges on back. Body mottled

green, usually with light stripe along spine, as well as *thin light line on tibia*. Belly uniform cream colour. Widely distributed and common in wet lowland areas. Breeding males give brief, nasal bray call while at the edge of water or propped up on aquatic vegetation. Small black-and-white eggs laid in shallow water.

DWARF RIDGED FROG
Ptychadena taenioscelis 25–40mm

A small, slender frog with long webbed hindlegs and rows of parallel skin ridges on the back. Tympanum smaller than the eye, and pale ridge runs under eye to neck. Back brown with

reticulate dark markings between three thin pale lines. A single continuous dark band passes from knee to knee up the thighs and below the vent. Found in savanna from southern Kenya to Angola and South Africa. Breeding male gives rapid chirp call as it sits in flooded grassland. Small groups of eggs laid in shallow water.

SHARP-NOSED RIDGED FROG
Ptychadena oxyrhynchus 50–85mm

Also known as the rocket frog, this medium-sized species has long, strong legs and can leap prodigious distances. Snout is sharp and toes extensively webbed. Brown-spotted back has several rows of *parallel skin ridges*, but

tibial and vertebral lines *absent*. Belly smooth and white, sometimes with yellowish groin. Inhabits savannas from South Africa to Senegal. In breeding season, males give sharp, extended trill call from the ground, usually near water. Up to 3,500 eggs are laid in short strings that later break apart and float on the water surface.

ORNATE FROG
Hildebrandtia ornata 50–70mm

Male and female mating

A small, fat, burrowing frog with a *smooth skin* and bright colour pattern. Hindfeet fat, *short-toed and feebly webbed*. Back with broad pale bands, separated by darker blotches, and thighs have dark bars. A dark band runs from snout, through eye to neck. Dusky throat has *pair of 'Y'-shaped light marks*. Males give harsh bellow from water's edge. Up to 1,200 eggs are laid in large floating masses in shallow marshy pans.

AFRICAN FROGS Family Pyxicephalidae

Most of these frogs were previously placed in the Ranidae. All are restricted to Africa, and form some of the most characteristic amphibians in sub-Saharan Africa.

PLIMPTON'S DAINTY FROG
Cacosternum plimptoni 15–25mm

A *minute, teardrop-shaped frog* with *narrow head* and flattened body. Limbs thin with *unwebbed* toes. Body smooth, varying in colour from green to brown, with spots or stripes. Belly cream and

has small grey or black spots. Lives in the grasslands of the Serengeti–Maasai Mara, sheltering under the ground in cracks or old rodent burrows in dry season. Breeds in flooded depressions, with breeding males calling while concealed in cracks or vegetation. Moves in small hops. Reluctant to enter water.

DWARF BULLFROG
Pyxicephalus edulis 80–120mm

A large, fat frog. *Skin rough with short folds and bumps.* Hindfeet fat, *short-toed and feebly webbed.* Lower jaw has *two sharp, teeth-like cusps.* Tympanum as large as the eye. Males have bright green backs, with darker spots

and often a pale vertebral stripe. Females duller olive colour. Inhabits coastal thicket and moist savanna. Over ¾ of the year spent dormant underground in cocoon of shed skin. Breeding males give *call like that of a barking dog* while sitting in flooded vegetation. Eggs are laid in shallow water and develop quickly. Frog is edible and eaten by local people. Larger inland forms (e.g. from Dodoma and the Kenya highlands) may be different species.

CRYPTIC SAND FROG
Tomopterna cryptotis 40–60mm

A small, stocky, burrowing frog that has a *glandular ridge* beneath the tympanum. Large, flattened *'spade' on the heel* allows it to dig rapidly backward into sandy soil. Back mottled and may comprise various shades of grey-tan to rusty brown. May have pale head patch or thin pale line down spine. Inhabits savannas from South to West Africa, and breeds in temporary ponds. Males sit on exposed mudbanks and give long call that sounds like telephone ringing: 'bing, bing, bing, bing, ...'. Eggs scattered singly in shallow water.

RED-BACKED SAND FROG
Tomopterna luganga 35–52mm

A small, stocky, burrowing frog that has a large, flattened *'spade' on heel* that allows it to dig rapidly backward into sandy soil. Tympanum is *oval*. Hindlimb *lacks* outer metatarsal. Back rusty brown with scattered tubercles. Has dark-edged pale patch behind eyes. Inhabits central arid savanna of Tanzania, where it breeds in temporary ponds.

RIVER FROGS
Amietia angolensis complex 60–90mm

A large frog with *long legs* and *strongly webbed toes*. Back olive, dark-spotted, with distinct folds on upper flanks. Belly light. Usually has dark mottling on throat. Inhabits permanent water bodies from Ethiopia to South Africa. Breeding males have swollen thumbs and dark nuptial pads, and call from water's edge, often partially submerged. The complex call consists of a long rattle that finishes with brief croak. Large numbers of eggs are laid in shallow water. Eats large insects and even other frogs.

DE WITTE'S FROG
Amietia wittei 40–60mm

A smallish, squat frog with large eyes and eardrums and *strongly webbed toes*. A prominent *black band runs from eye* onto side of blotched brown body. Lower flanks mottled with yellow and cream. Prefers marshy areas and stream margins in montane grasslands and is active during the day, shuffling down into marsh vegetation when disturbed. Found in scattered populations from Ethiopia to the eastern Congo region. Eggs are laid in water and the black tadpoles develop slowly.

TYPICAL FROGS Family Ranidae

This family once included many African frogs, but is now mainly restricted to northern hemisphere, with only a few African representatives.

GOLDEN-BACKED FROG
Hylarana galamensis 50–86mm

An attractive frog with strong limbs and *smooth skin*. Eyes situated on *side of head*, and tympanum *slightly smaller than eye*. Toes *half-webbed*, but tips are *not expanded*. A broad golden band down back gives this frog its common name. Flanks dark brown, usually with lighter patches. Bold cream line runs from lips to base of hindlimb. Inhabits permanent ponds in coastal bush and moist savanna. Breeding occurs in still pools and the eggs float on the water surface.

FORK-TONGUED FROGS Family Discoglossidae

A group of mainly Asian frogs that possess forked tongues. There is just one African species.

CROWNED BULLFROG
Hoplobatrachus occipitalis 80–120mm

A large, squat frog. *Skin rough with short folds and bumps*. Eyes *on top* of head and 'crowned' by a *fold of skin* that runs between backs of eyes. Has *large tympanum* and eye has *diamond-shaped pupil*. Toes *strongly webbed*. Body colour varies from dark brown to

green, usually with some white spots. Belly and sometimes flanks are yellow. Prefers open woodland. Its presence in forests a sign of disturbance. Mainly found in Great Lakes region, but may be expanding its range. Tadpoles are carnivorous with horny 'teeth' to rasp other tadpoles, particular those of ridged frogs (pages 147–8)

CAECILIANS Order Gymnophiona

Bizarre legless amphibians restricted to tropical regions of South America, Africa and Asia. Fertilisation is internal and they lay a few eggs in moist soil. In some species these undergo direct development.

KIRK'S CAECILIAN
Scolecomorphus kirkii 200–270mm

An *elongate, snake-like* amphibian that *lacks legs* and burrows underground. Eyes situated at the *base of small tentacles,* and may be extruded like the eye of a snail. Skin smooth with numerous grooves that give it the appearance of a rubbery worm. Has between *129 and 148 body annuli,* and vent is *longitudinal.* Burrows underground in moist savanna. Little known of its reproduction. Probably gives birth to small number of babies.

ULUGURU PINK CAECILIAN
Boulengerula uluguruensis 200–270mm

Very thin, with *no visible eyes, pointed snout, underslung mouth,* and small tentacle near to corner of mouth. Has 132–148 body annuli and vent is *transverse.* Bright pink above and below. Little known about its life, but probably feeds on soil invertebrates such as termites, and female lays eggs. Common in moist forest soils of the Ulugurus.

ILLUSTRATED GLOSSARY

Amphibian external features

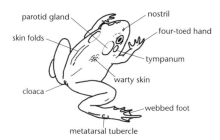

- parotid gland
- nostril
- skin folds
- four-toed hand
- tympanum
- warty skin
- cloaca
- webbed foot
- metatarsal tubercle

Lizard colour patterns

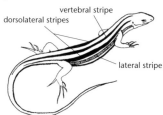

- vertebral stripe
- dorsolateral stripes
- lateral stripe

Chelonian external features

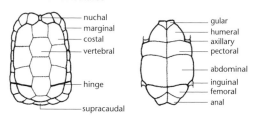

- nuchal
- marginal
- costal
- vertebral
- hinge
- supracaudal
- gular
- humeral
- axillary
- pectoral
- abdominal
- inguinal
- femoral
- anal

Head scales of a colubrid snake

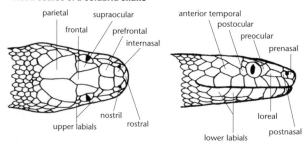

- parietal
- supraocular
- frontal
- prefrontal
- internasal
- nostril
- upper labials
- rostral
- anterior temporal
- postocular
- preocular
- prenasal
- loreal
- postnasal
- lower labials

GLOSSARY OF TERMS

Amplexus: The mating clasp of amphibians.

Apical pits: Minute sense organs on body scales of snakes.

Arboreal: Living in or among trees.

Bridge: The side of a chelonian shell where the carapace joins the plastron.

Carapace: The upper surface of a chelonian shell.

Caudal: Pertaining to the tail region.

Chelonian: A shield reptile (tortoises, terrapins and turtles).

Clutch: All the eggs laid by a female at one time.

Cryptic: Hidden or camouflaged.

Diurnal: Active during the day.

Dorsal: Pertaining to the upper surface of the body.

Dorsolateral: Pertaining to the upper flanks; that part of the body bordering the backbone.

Ectotherm: An animal, including all reptiles, that obtains its body heat externally, usually from basking in the sun.

Femoral: Pertaining to the upper part (thigh) of the hindlimb.

Granular: Small, usually non-overlapping, scales.

Gular: Pertaining to the throat region; a plate on the plastron of a chelonian shell.

Hemipenes: Paired sex organs of male squamates that are stored in the tail base (singular: Hemipenis).

Hinge: A flexible joint in the shell of some chelonians.

Keel: A prominent ridge occurring on the back of some chelonians and on the scales of some lizards and snakes.

Lamellae: Organ structures arranged in parallel rows.

Marginals: Scutes around the edge of the chelonian carapace.

Nocturnal: Active at night.

Nuchal: A scute at the front of the carapace of a chelonian shell.

Occipital: Pertaining to the back of the skull.

Osteoderm: A body scale containing a bony layer.

Parotid glands: Large glands that produce toxic secretions, located on each side of the neck in toads.

Plastron: The lower surface of a chelonian shell.

Recurved: Descriptive of a tooth that bends backwards.

Rostral: Pertaining to the rostrum (nose); a scale at the front of the nose of a reptile.

Scansor: Specialised scales found on the toe-tips of many geckos. They are covered in millions of minute setae ('hairs') that allow the gecko to climb vertical surfaces.

Scute: An enlarged horny plate on a chelonian shell.

Squamate: A scaled reptile: snakes, lizards and worm lizards.

Subcaudal: Beneath the tail region.

Temporal: Pertaining to the side of the head behind the eye.

Tubercle: An enlarged scale on the body of a lizard.

Tympanum: The eardrum; usually exposed in amphibians.

Ventral: Pertaining to the lower surface of the body.

Vestigial: Being of a simpler structure than in an ancestor.

FURTHER READING

Channing, A and Howell, K M. 2005. *Amphibians of East Africa*. Cornell University Press.

Schiotz, A. 1999. Treefrogs of Africa. Edition Chimaira, Frankfurt-am-Main.

Spawls, S, Howell, K, Drewes, R and Ashe, J. 2002. *A Field Guide to the Reptiles of East Africa*. Academic Press, London.

Spawls, S. and Mathews, G. 2013. *Kenya, A Natural History*. T & A.D. Poyser, London.

Tilbury, C. 2010. *Chameleons of Africa – An Atlas*. Edition Chimaira, Frankfurt am Main, Germany.

USEFUL WEBSITES

Bwong, B, Malonza, P, Muchai, V, Spawls, S and Wasonga,V. *Kenya Reptile Atlas*. **http://www.kenyareptileatlas.com/**

East African Snakes and Other Reptiles group. **https://www.facebook.com/groups/662521540444058/**

INDEX

SNAKES

LIZARDS

CROCODILES